# An Illustrative Guide to Multivariable and Vector Calculus

Stanley J. Miklavcic

# An Illustrative Guide to Multivariable and Vector Calculus

In collaboration with Ross A. Frick

 Springer

Stanley J. Miklavcic
University of South Australia
(Mawson Lakes Campus)
Adelaide, SA, Australia

ISBN 978-3-030-33461-1          ISBN 978-3-030-33459-8    (eBook)
https://doi.org/10.1007/978-3-030-33459-8

Mathematics Subject Classification (2010): 26B05, 26B10, 26B12, 26B15, 26B20

This Springer imprint is published by the registered company Springer Nature Switzerland AG
The registered company address is: Gewerbestrasse 11, 6330 Cham, Switzerland

# Dedication

To my children Arya, Nadia, Jacob, and David.

# Preface

This book originated as a set of lectures prepared for courses given by me at the University of Linköping in Sweden and at the University of South Australia in Australia. At Linköping University the material (apart from Section 3.E) was delivered in a second year, single semester course (14 weeks, 2 two-hour lectures per week) to engineering students, with the first half focused on the differential calculus of real-valued multivariable functions, while the second half was divided between integral calculus and vector calculus. At the University of South Australia the subject was delivered in two separate semester courses (12 weeks, 2 two-hour lectures per week), the first of which was offered to second year engineering, science and mathematics students and featured differential and integral calculus, including an introduction to partial differential equations. The second course, taken mostly by third year mathematics and science students, dealt with vector calculus, although only the first five weeks of that course was covered by the material in this book.

The lectures generally were so well-received by students that it was thought the material might appeal to a wider audience. Having taken the decision to convert my notes into a book, I aimed for a document of manageable size rather than generate yet another bulky tome on calculus. The result is a book that students can carry easily to and from class, can take out and leaf through on the library lawn, or in a booth of a pub, or while lying on the banks of a river waiting for the fish to bite.

Very many ideas in mathematics are more readily conveyed and more easily appreciated when presented visually. This is certainly true of multivariable and vector calculus, and as my lecture material took advantage of many visual devices, I sought to capture the spirit if not the body of these same devices in printed form. Consequently, the majority of concepts are introduced and explained with the support of figures and graphics as well as the generous use of colour. Indeed, colour is used to highlight specific pieces of information, to emphasize relationships between variables in different equations, and to distinguish different roles or actions. The inevitable issue of colour blindness was raised in the course of the book's development. To minimize difficulties, colour typesetting has been configured to allow for some degree of differentiation even by those readers with impaired colour vision. In addition, colour has been implemented so as only to benefit one's understanding, and *not* as an essential condition for understanding.

The book is self-contained and complete as an introduction to the theory of the differential and integral calculus of both real-valued and vector-valued multivariable functions. The entire material is suitable as a textbook in its own right for one to two, semester-long courses in either the second year or third year of University studies, and for students who have already completed courses in single variable calculus and linear algebra. Some selection of content may be necessary depending on student need and time available. For instance, as the topic of partial differential equations (PDEs) is normally offered as a separate course to mathematics students, Section 3.E can be skipped in a multivariable calculus course. On the other hand, a course in PDEs is not always included in engineering and science curricula, so Section 3.E is a pragmatic, albeit brief, introduction to the subject, particularly as its focus is on solving PDEs in simple cases. Alternatively, because of its illustrative emphasis, the book can also perform the role of a reference text to complement one of the more standard textbooks in advanced calculus, such as [1], thus providing the student with a different visual perspective.

Consequent to the ambition of producing a portable book, the reader should not be surprised that some areas of the calculus are not covered in detail. One other notable sacrifice is mathematical rigour. There are very few proofs included and those that have been are deliberately sketchy, included only to give students a rational justification for, or to illustrate the origin of, an idea. Consequently, students of pure mathematics may want to complement this book with one that offers a deeper analysis, such as [2].

Within each chapter is a sequence of Mastery Checks, exercises on the topic under discussion that are usually preceded by solved examples. Students are encouraged to attempt these Mastery Checks and keep a record of their solutions for future reference. To reinforce the ideas, additional exercises appear at the end of each chapter to supplement the Mastery Checks. Solutions to both sets of exercises are available to instructors upon request. I have limited the number of problems in order to restrict the size of the book, assuming that students would have access to auxiliary exercises in more standard treatises. All the same, the book contains over 90 Mastery Checks and over 120 Supplementary Exercises, many with multiple parts.

The reader should be aware that I have made use of mathematical symbols (such as $\implies$ and $\exists$) and abbreviations (w.r.t., 3D) in place of text, a common practice in mathematics texts and research literature. A glossary of definitions can be found at the end of the book. Wherever they appear in the book they should be read as the pieces of text they replace. Finally, for easy reference a list of Important Formulae, covering various topics in multivariable and vector calculus, is given on page xiii.

## Acknowledgements

In drafting this book I had great pleasure in working closely with my colleague Ross Frick who was instrumental in turning my original lecture material and supplementary notes into book form. His skill with LaTeX and MATLAB® was critical in this endeavour. I would also like to thank Dr. Loretta Bartolini, Mathematics Editor at Springer, for her strong support and encouragement of this venture and for her efficient handling of the publication of this book.

I will forever be indebted to Julie for her patience and enduring support over the many, many months of editing and re-editing to which this book was subjected. It is no exaggeration to say that without her understanding the task of completing this book would have been a far greater challenge than it has been.

Lastly, I would like to thank the students who have taken my course over the years, particularly those (now graduate) students who gave feedback on the notes prior to their publication. Their general enthusiasm has been an absolutely essential factor in getting the book to this point. I hope that future students of this important area of mathematics will also enjoy and be inspired by what this little volume has to offer.

Adelaide, Australia                                   Stanley J. Miklavcic
December 2019

# Contents

# Important Formulæ

## Multivariable calculus

- Suppose $f, g, h \in C^1$ and $w = f(u, v)$ where $u = g(x, y)$ and $v = h(x, y)$, then the partial derivative of $w$ with respect to $x$ is given by $\frac{\partial w}{\partial x} = \frac{\partial f}{\partial u}\frac{\partial g}{\partial x} + \frac{\partial f}{\partial v}\frac{\partial h}{\partial x}$, and similarly for the partial derivative of $w$ with respect to $y$.

- Suppose $f \in C^3$ at the point $(a, b)$, then for points $(x, y)$ in a neighbourhood of $(a, b)$ the function $f$ has the following Taylor approximation:

$$f(x, y) = f(a, b) + f_x(a, b)\Delta x + f_y(a, b)\Delta y + \frac{1}{2}Q(\Delta x, \Delta y) + R(\Delta x, \Delta y)$$

where

$$Q(\Delta x, \Delta y) = f_{xx}(a, b)(\Delta x)^2 + 2f_{xy}(a, b)\Delta x \Delta y + f_{yy}(a, b)(\Delta y)^2$$

and $R$ is a remainder term of order $((\Delta x^2 + \Delta y^2)^{3/2})$.

- $f = f(x, y, z) \in C^2$ is a solution of Laplace's equation in domain $D \subset \mathbb{R}^3$ if $f$ satisfies

$$\nabla^2 f = \frac{\partial^2 f}{\partial x^2} + \frac{\partial^2 f}{\partial y^2} + \frac{\partial^2 f}{\partial z^2} = 0.$$

- The Jacobian for the transformation $(x, y) \mapsto (u, v)$:

$$J = \frac{\partial(x, y)}{\partial(u, v)} = \begin{vmatrix} \frac{\partial x}{\partial u} & \frac{\partial y}{\partial u} \\ \frac{\partial x}{\partial v} & \frac{\partial y}{\partial v} \end{vmatrix}$$

- Orthogonal curvilinear coordinate systems and their corresponding change of variables:

  1. Polar coordinates: $r \geq 0$ and $0 \leq \theta \leq 2\pi$; Jacobian, $J = r$.

  $$\{\, x = r\cos\theta,\ y = r\sin\theta \,\}$$

  2. Cylindrical polar coordinates: $r \geq 0$, $0 \leq \theta \leq 2\pi$ and $z \in \mathbb{R}$; Jacobian, $J = r$.

  $$\{\, x = r\cos\theta,\ y = r\sin\theta,\ z = z \,\}$$

  3. Spherical polar coordinates: $\rho \geq 0$, $0 \leq \phi \leq \pi$, $0 \leq \theta \leq 2\pi$; Jacobian, $J = \rho^2 \sin\phi$.

  $$\{\, x = \rho\sin\phi\cos\theta,\ y = \rho\sin\phi\sin\theta,\ z = \rho\cos\phi \,\}$$

- A level set to a function $f(x,y)$ is the set $\{(x,y) : \text{s.t. } f(x,y) = C\}$ for some real constant $C$. The gradient of $f$ is always normal to a level set of $f$.

- An iterated integral of a function of two variables over a $y$-simple domain $D = \{(x,y) : a \leq x \leq b; g_1(x) \leq y \leq g_2(x)\}$

$$\iint_D f(x,y)\,\mathrm{d}A = \int_a^b \mathrm{d}x \int_{g_1(x)}^{g_2(x)} f(x,y)\,\mathrm{d}y.$$

- An iterated integral of a function of two variables over a $x$-simple domain $D = \{(x,y) : c \leq y \leq d; h_1(y) \leq x \leq h_2(y)\}$

$$\iint_D f(x,y)\,\mathrm{d}A = \int_c^d \mathrm{d}y \int_{h_1(y)}^{h_2(y)} f(x,y)\,\mathrm{d}x.$$

- For a bijective transformation $D \ni (x,y) \mapsto (u,v) \in E$ with Jacobian determinant, $J \neq 0$, the double integral of $f(x,y) = F(u,v)$ is

$$\iint_D f(x,y)\,\mathrm{d}x\mathrm{d}y = \iint_E F(u,v)\left|\frac{\partial(x,y)}{\partial(u,v)}\right|\mathrm{d}u\mathrm{d}v.$$

## Vector calculus

- A $C^1$ vector field $\mathbf{F} = \mathbf{F}(\mathbf{x})$ defined in some domain $D \subset \mathbb{R}^3$ is said to be

  1. solenoidal in $D$, if $\nabla \cdot \mathbf{F} = 0$ in $D$;
  2. irrotational in $D$, if $\nabla \times \mathbf{F} = \mathbf{0}$ in $D$;
  3. conservative in $D$, if $\mathbf{F} = \nabla \phi$ in $D$ for some $C^2$ real-valued function $\phi = \phi(\mathbf{x})$.

- In terms of a 3D curvilinear coordinate system $\{\xi_1, \xi_2, \xi_3\}$, with unit vectors $\mathbf{a}_1, \mathbf{a}_2, \mathbf{a}_3$, and scale factors $h_1, h_2, h_3$, the gradient, divergence and curl operations on scalar ($\phi \in C^1$) and vector ($\mathbf{F} \in C^1$) fields, respectively, take the form

$$\nabla \phi = \frac{1}{h_1} \frac{\partial \phi}{\partial \xi_1} \mathbf{a}_1 + \frac{1}{h_2} \frac{\partial \phi}{\partial \xi_2} \mathbf{a}_2 + \frac{1}{h_3} \frac{\partial \phi}{\partial \xi_3} \mathbf{a}_3$$

$$\nabla \cdot \mathbf{F} = \frac{1}{h_1 h_2 h_3} \left[ \frac{\partial}{\partial \xi_1} (h_2 h_3 F_1) + \frac{\partial}{\partial \xi_2} (h_1 h_3 F_2) + \frac{\partial}{\partial \xi_3} (h_1 h_2 F_3) \right],$$

$$\nabla \times \mathbf{F} = \frac{1}{h_1 h_2 h_3} \begin{vmatrix} h_1 \mathbf{a}_1 & h_2 \mathbf{a}_2 & h_3 \mathbf{a}_3 \\ \frac{\partial}{\partial \xi_1} & \frac{\partial}{\partial \xi_2} & \frac{\partial}{\partial \xi_3} \\ h_1 F_1 & h_2 F_2 & h_3 F_3 \end{vmatrix}$$

For Cartesian coordinates $\{\xi_1, \xi_2, \xi_3\} = \{x_1, x_2, x_3\} = \{x, y, z\}$, $h_1 = h_2 = h_3 = 1$, and $(\mathbf{a}_1 = \mathbf{e}_1 = \mathbf{i}; \mathbf{a}_2 = \mathbf{e}_2 = \mathbf{j}; \mathbf{a}_3 = \mathbf{e}_3 = \mathbf{k})$.

- Some useful vector identities. Suppose $\psi, \phi : R^3 \to R$ and $h : R \to R$ are $C^1$ scalar-valued functions, $\mathbf{f}, \mathbf{g} : R^3 \to R^3$ are $C^1$ vector-valued functions, $\mathbf{x} = (x, y, z)$ is a position vector of length $r = |\mathbf{x}| = \sqrt{x^2 + y^2 + z^2}$ and $\mathbf{c}$ is a constant vector.

(1)  $\nabla(\phi \psi) = \psi \nabla \phi + \phi \nabla \psi$

(2)  $\nabla \cdot (\phi \mathbf{f}) = \phi \nabla \cdot \mathbf{f} + \mathbf{f} \cdot \nabla \phi$

(3)  $\nabla \times (\phi \mathbf{f}) = \phi \nabla \times \mathbf{f} + \nabla \phi \times \mathbf{f}$

(4)  $\nabla(\mathbf{f} \cdot \mathbf{g}) = (\mathbf{f} \cdot \nabla)\mathbf{g} + (\mathbf{g} \cdot \nabla)\mathbf{f} + \mathbf{f} \times (\nabla \times \mathbf{g}) + \mathbf{g} \times (\nabla \times \mathbf{f})$

(5)  $\nabla \cdot (\mathbf{f} \times \mathbf{g}) = \mathbf{g} \cdot (\nabla \times \mathbf{f}) - \mathbf{f} \cdot (\nabla \times \mathbf{g})$

(6)  $\nabla \times (\mathbf{f} \times \mathbf{g}) = \mathbf{f}(\nabla \cdot \mathbf{g}) - \mathbf{g}(\nabla \cdot \mathbf{f}) + (\mathbf{g} \cdot \nabla)\mathbf{f} - (\mathbf{f} \cdot \nabla)\mathbf{g}$

(7)  $\nabla \times (\nabla \phi) = \mathbf{0}$

(8)  $\nabla \cdot (\nabla \times \mathbf{f}) = 0$

(9)  $\nabla \times (\nabla \times \mathbf{f}) = \nabla(\nabla \cdot \mathbf{f}) - \nabla^2 \mathbf{f}$

(10)  $\nabla \cdot \mathbf{x} = 3$

(11)  $\nabla h(r) = \dfrac{dh}{dr} \dfrac{\mathbf{x}}{r}$

(12)  $\nabla \cdot (h(r)\mathbf{x}) = 3h(r) + r\dfrac{dh}{dr}$

(13)  $\nabla \times (h(r)\mathbf{x}) = \mathbf{0}$

(14)  $\nabla(\mathbf{c} \cdot \mathbf{x}) = \mathbf{c}$

(15)  $\nabla \cdot (\mathbf{c} \times \mathbf{x}) = 0$

(16)  $\nabla \times (\mathbf{c} \times \mathbf{x}) = 2\mathbf{c}$

- Vector integration

    1. Line integral of $\mathbf{f} = \mathbf{f}(\mathbf{r})$ over $\Gamma = \{\mathbf{r} = \mathbf{r}(t) : a \le t \le b\}$:

    $$\int_\Gamma \mathbf{f} \cdot d\mathbf{r} = \int_a^b \mathbf{f}(\mathbf{r}(t)) \cdot \frac{d\mathbf{r}}{dt} dt$$

    2. Surface integral of $\mathbf{f} = \mathbf{f}(\mathbf{r})$ over $S = \{\mathbf{r} = \mathbf{r}(u,v) : (u,v) \in D \subset \mathbb{R}^2\}$ with unit surface normal $\mathbf{N}$:

    $$\iint_S \mathbf{f} \cdot d\mathbf{S} = \iint_S \mathbf{f} \cdot \mathbf{N} \, dS = \iint_D \mathbf{f}(\mathbf{r}(u,v)) \cdot \left(\frac{\partial \mathbf{r}}{\partial u} \times \frac{\partial \mathbf{r}}{\partial v}\right) du \, dv$$

    3. Green's theorem for $\mathbf{f} = (f_1, f_2) \in C^1$ over a finite 2D region $D$ bounded by a positively oriented closed curve $\Gamma$:

    $$\oint_\Gamma \mathbf{f} \cdot d\mathbf{r} = \oint_\Gamma (f_1 \, dx + f_2 \, dy) = \iint_D \left(\frac{\partial f_2}{\partial x} - \frac{\partial f_1}{\partial y}\right) dA$$

    4. Gauss's theorem (divergence theorem) for $\mathbf{f} \in C^1$ over a finite 3D region $V$ bounded by smooth closed surface $S$ with outward pointing, unit surface normal, $\mathbf{N}$:

    $$\iiint_V (\nabla \cdot \mathbf{f}) dV = \oiint_S \mathbf{f} \cdot \mathbf{N} \, dS$$

    5. Stokes's theorem for $\mathbf{f} \in C^1$ defined on a smooth surface $S$ with unit surface normal $\mathbf{N}$ and bounded by a positively oriented, closed curve $\Gamma$:

    $$\iint_S (\nabla \times \mathbf{f}) \cdot d\mathbf{S} = \iint_S (\nabla \times \mathbf{f}) \cdot \mathbf{N} \, dS = \oint_\Gamma \mathbf{f} \cdot d\mathbf{r}$$

# Chapter 1

# Vectors and functions

Many mathematical properties possessed by functions of several variables are couched in geometric terms and with reference to elementary set theory. In this introductory chapter I will revisit some of the concepts that will be needed in later chapters. For example, vector calculus springs naturally from vector algebra so it is appropriate to begin the review with the latter topic. This is followed by a short review of elementary set theory, which will be referred to throughout the book and will indeed help establish many foundation concepts in both the differential and integral calculus. Coordinate systems and the notion of level sets are also discussed. Once again, both topics find application in differential and integral multivariable calculus, as well as in vector calculus.

It goes without saying that a review of single-variable functions is helpful. This begins in this chapter (Section 1.C), but continues in Chapters 2, 3 and 4 as needed.

To help appreciate the behaviour of multivariable functions defined on two or higher dimensional domains, It is useful to at least visualize their domains of definition. Sometimes, though, it is possible, as well as necessary, to visualize the entire graph of a function, or some approximation to it. Some people are more hard-wired to visual cues and visual information, while others are more comfortable with abstract ideas. Whatever your preference, being able to draw figures is always useful. Consequently, in this chapter we also review some basic 3D structures and show how to draw them using MATLAB®. Of course, other software will serve equally well. In the event of the reader being unable to access software solutions, there is included a subsection which may hopefully illustrate, by example, how one can obtain a picture of a region

© Springer Nature Switzerland AG 2020
S. J. Miklavcic, *An Illustrative Guide to Multivariable and Vector Calculus*,
https://doi.org/10.1007/978-3-030-33459-8_1

or of a function graph directly from a mathematical formula or equation. Although it is not possible to offer a general procedure that works in all cases, some of the steps may be applicable in other instances.

# 1.A   Some vector algebra essentials

**Unit vectors in 3-space.**
Let $a > 0$ be a scalar, and let

$$
\begin{aligned}
\boldsymbol{v} &= (\alpha, \beta, \gamma) \\
&= \alpha \mathbf{i} + \beta \mathbf{j} + \gamma \mathbf{k} \\
&= \alpha \mathbf{e}_1 + \beta \mathbf{e}_2 + \gamma \mathbf{e}_3
\end{aligned}
$$

be a vector in $\mathbb{R}^3$ (see Section 1.B) with $x$-, $y$-, and $z$-components $\alpha$, $\beta$, and $\gamma$.

This vector has been written in the three most common forms appearing in current texts. The sets $\{\mathbf{i}, \mathbf{j}, \mathbf{k}\}$ and $\{\mathbf{e}_1, \mathbf{e}_2, \mathbf{e}_3\}$ represent the same set of unit vectors in mutually orthogonal directions in $\mathbb{R}^3$. The first form simply shows the components along the three orthogonal directions without reference to the unit vectors themselves, although the unit vectors and the coordinate system are implicit in this notation. The reader should be aware that we shall have occasion to refer to vectors using any of the three formats. The choice will depend on what is most convenient at that time without compromising understanding.

Multiplying a vector $\boldsymbol{v}$ with a scalar will return a new vector with either the same direction if the scalar is positive or the opposite direction if the scalar is negative. In either case the resulting vector has different magnitude (Figure 1.1). This re-scaling will be a feature in Chapter 5 where we will need vectors of unit magnitude. For $a\boldsymbol{v}$, with $a \in \mathbb{R}$, to be a unit vector we must have

$$
|a\boldsymbol{v}| = |a||\boldsymbol{v}| = a\sqrt{\alpha^2 + \beta^2 + \gamma^2} = 1, \ i.e., \ a = \frac{1}{\sqrt{\alpha^2 + \beta^2 + \gamma^2}}.
$$

Therefore, to construct a unit vector in the direction of a specific vector $\boldsymbol{v}$ we simply divide $\boldsymbol{v}$ by its length:

$$
\boldsymbol{N} = \frac{\boldsymbol{v}}{|\boldsymbol{v}|}.
$$

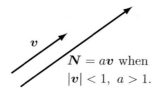

**Figure 1.1**  The unit vector.

**The product of two vectors in 3-space.**
Let $u$ and $v$ be two non-parallel vectors in $\mathbb{R}^3$:

$$u = (a_1, a_2, a_3) \quad v = (b_1, b_2, b_3).$$

There are two particular product operations that we will utilize on many occasions. These are the vector and scalar products. From them very useful information can be extracted.

(a) A vector *perpendicular* to *both* $u$ and $v$ is

$$w = u \times v = \begin{vmatrix} \mathbf{i} & \mathbf{j} & \mathbf{k} \\ a_1 & a_2 & a_3 \\ b_1 & b_2 & b_3 \end{vmatrix}$$
$$= (a_2 b_3 - a_3 b_2, a_3 b_1 - a_1 b_3, a_1 b_2 - a_2 b_1)$$
$$= -v \times u.$$

This is called the "vector" or "cross" product. Note that $u \times v$ is antiparallel to $v \times u$. The relationship between the three vectors is shown in Figure 1.5.

(b) The magnitude of the vector (cross) product of two vectors

$$|u \times v| = |w| = \left| \begin{vmatrix} \mathbf{i} & \mathbf{j} & \mathbf{k} \\ a_1 & a_2 & a_3 \\ b_1 & b_2 & b_3 \end{vmatrix} \right| = \sqrt{(a_2 b_3 - a_3 b_2)^2 + \cdots}$$

gives the *area* of a plane parallelogram whose side lengths are $|u|$ and $|v|$ (Figure 1.2).

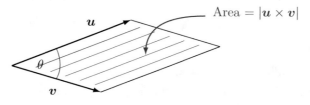

**Figure 1.2**  The $u \times v$ parallelogram.

The magnitude of the vector product is then given by

$$|\boldsymbol{w}| = |\boldsymbol{u}||\boldsymbol{v}|\sin\theta,$$

where $\theta$ is the angle between $\boldsymbol{u}$ and $\boldsymbol{v}$ lying in the plane defined by $\boldsymbol{u}$ and $\boldsymbol{v}$.

(c) The angle $\theta$ between the vectors $\boldsymbol{u}$ and $\boldsymbol{v}$ (Figure 1.3) can also be (and usually is) obtained from the "scalar" or "dot" product, defined as

$$\boldsymbol{u} \cdot \boldsymbol{v} = a_1 b_1 + a_2 b_2 + a_3 b_3$$
$$= |\boldsymbol{u}||\boldsymbol{v}|\cos\theta.$$

So we have

$$\cos\theta = \frac{\boldsymbol{u} \cdot \boldsymbol{v}}{|\boldsymbol{u}||\boldsymbol{v}|}.$$

If $\theta = 0$, then the vectors are parallel and $\boldsymbol{u} \cdot \boldsymbol{v} = |\boldsymbol{u}||\boldsymbol{v}|$. If $\theta = \dfrac{\pi}{2}$, then the vectors are orthogonal and $\boldsymbol{u} \cdot \boldsymbol{v} = 0$. For example, in 2(a) above, $\boldsymbol{w} \cdot \boldsymbol{u} = \boldsymbol{w} \cdot \boldsymbol{v} = 0$ as $\boldsymbol{w}$ is orthogonal to both $\boldsymbol{u}$ and $\boldsymbol{v}$.

We will make extensive use of these products in Chapter 2 (Sections 2.E and 2.G) and throughout Chapter 5.

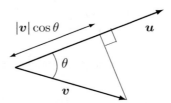

**Figure 1.3** The projection of $\boldsymbol{v}$ onto $\boldsymbol{u}$.

**A plane in 3-space.**
The equation of a plane in $\mathbb{R}^3$, expressed mathematically as

$$P = \{(x, y, z) : ax + by + cz = d; \ a, b, c \text{ not all zero.}\}$$

can be determined knowing either

(a) three non-collinear points on the plane; or

(b) one point and two non-parallel vectors lying parallel to the plane.

Consider alternative 3(a). Let $\boldsymbol{x}_i = (x_i, y_i, z_i)$, $i = 1, 2, 3$, be the three points. Construct two vectors, $\boldsymbol{u}$ and $\boldsymbol{v}$ in the plane (Figure 1.4):

$$u = x_2 - x_1 = \left( x_2 - x_1, y_2 - y_1, z_2 - z_1 \right)$$
$$v = x_3 - x_1 = \left( x_3 - x_1, y_3 - y_1, z_3 - z_1 \right)$$

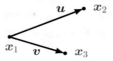

**Figure 1.4**  Construction of vectors $u$ and $v$.

As long as $x_1$, $x_2$ and $x_3$ are not collinear, then $u$ and $v$ will not be super-imposed or even parallel and

$$w = u \times v = (\alpha, \beta, \gamma)$$

will be a vector normal (perpendicular) to $u$ and $v$ and thus normal to the plane in which the $x_i$ lie.

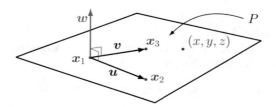

**Figure 1.5**  Construction of the plane $P$.

By convention, the direction of $w = u \times v$ is given by the **right-hand rule**: *Using your right hand, point the index finger in the direction of $u$ and the middle finger in the direction of $v$, then the thumb shows the direction of $w$.* The vector product and its various geometric properties will play central roles in Sections 2.C and 5.D.

Now consider alternative 3(b):
Let $(x, y, z)$ be any point in the plane $P$ except for the given points $(x_i, y_i, z_i)$, $\quad i = 1, 2, 3$. We construct the plane vector $(x - x_1, y - y_1, z - z_1)$ that joins this new point with the point $x_1$. Using concept 2(c) we have

$$w \cdot \left( x - x_1, y - y_1, z - z_1 \right) = 0$$
$$\implies \quad \alpha(x - x_1) + \beta(y - y_1) + \gamma(z - z_1) = 0$$
$$\implies \quad \alpha x + \beta y + \gamma z = \mathcal{K}.$$

The scalar product is thus instrumental in giving us the equation for the plane $P$ with $a = \alpha$, $b = \beta$, $c = \gamma$, and $d = \mathcal{K}$.

Alternative 3(b) is actually a version of alternative 3(a), except we are here given $\boldsymbol{u}$ and $\boldsymbol{v}$ with which to create the orthogonal vector $\boldsymbol{w}$. This method of deriving the equation of a plane will be utilized in Sections 2.C and 5.D.

**A line in 3-space.**
The general equation of a line,

$$L = \left\{ (x, y, z) : \frac{x - x_0}{a} = \frac{y - y_0}{b} = \frac{z - z_0}{c} \right\}, \tag{1.1}$$

can be derived using analogous reasoning. We need to know either

(a) two points on the line $L$, or

(b) one point and one vector parallel to $L$.

Consider alternative 4(a).
Let $(x_i, y_i, z_i)$, $i = 0, 1$, be the two given points. Construct the vector $\boldsymbol{u}$ directed from one point to the other:

$$\boldsymbol{u} = (x_1 - x_0, y_1 - y_0, z_1 - z_0) = (\alpha, \beta, \gamma).$$

As the two points line in the straight line $L$ so too must the vector $\boldsymbol{u}$ as shown in Figure 1.6. Note that as in Point 3, this construction leads directly to alternative 4(b) where $\boldsymbol{u}$ is given.

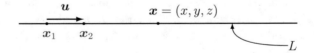

**Figure 1.6**  Vector $\boldsymbol{u}$ parallel to line $L$.

Using either alternative set of information, any point $(x, y, z)$ on $L$ can be determined by simple vector addition,

$$\boldsymbol{x} = \boldsymbol{x}_0 + t\,\boldsymbol{u}, \qquad t \in \mathbb{R}. \tag{1.2}$$

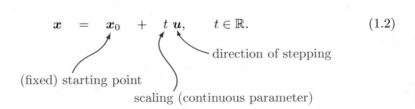

As indicated, this equation says that to determine any new point on the line we begin at a convenient starting point $\boldsymbol{x}_0$ and continue in the direction $\boldsymbol{u}$ a distance determined by the scalar $t$.

This simple vector equation is equivalent to the general equation given in Equation (1.1) above. We get the latter by splitting Equation (1.2) into its components and solving each component equation for the common scalar variable $t$.

A particularly important feature of this equation, which is linear in the parameter $t$, emerges from the single-variable derivative of each component

$$\frac{\mathrm{d}x}{\mathrm{d}t} = \alpha, \; \frac{\mathrm{d}y}{\mathrm{d}t} = \beta, \; \frac{\mathrm{d}z}{\mathrm{d}t} = \gamma.$$

Combining these into a vector equation we have that

$$\frac{\mathrm{d}\boldsymbol{x}}{\mathrm{d}t} = \boldsymbol{u} \quad \text{— the tangent vector.}$$

This last result is elementary, but has important applications in Sections 5.A and 5.C, where the straight line concepts are generalized to the case of nonlinear curves.

**The scalar triple product $\boldsymbol{u} \cdot (\boldsymbol{v} \times \boldsymbol{w})$.**
Let $\boldsymbol{u}, \boldsymbol{v}, \boldsymbol{w}$ be three non-parallel vectors. These define the edges of a parallelepiped as shown in Figure 1.7.

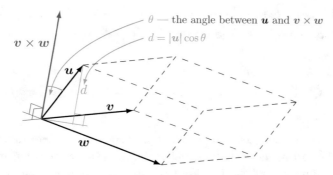

**Figure 1.7**  The $\boldsymbol{u}$, $\boldsymbol{v}$, $\boldsymbol{w}$ parallelepiped.

To form the scalar triple product, we first form the vector product of $\boldsymbol{v}$ and $\boldsymbol{w}$, $\boldsymbol{v} \times \boldsymbol{w}$, and then form the scalar product of that result and $\boldsymbol{u}$. The magnitude of the scalar triple product, which is found using Point 2(a), is given by

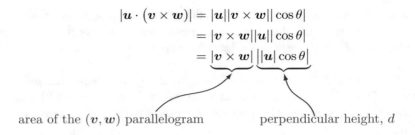

$$|u \cdot (v \times w)| = |u||v \times w|| \cos \theta|$$
$$= |v \times w||u|| \cos \theta|$$
$$= \underbrace{|v \times w|}\,\underbrace{||u| \cos \theta|}$$

area of the $(v, w)$ parallelogram          perpendicular height, $d$

This gives the *volume of the parallelepiped* formed by the vectors $u$, $v$, and $w$. Let

$$u = (a_1, a_2, a_3) = a_1 \mathbf{i} + a_2 \mathbf{j} + a_3 \mathbf{k},$$
$$v = (b_1, b_2, b_3) = b_1 \mathbf{i} + b_2 \mathbf{j} + b_3 \mathbf{k}, \text{ and}$$
$$w = (c_1, c_2, c_3) = c_1 \mathbf{i} + c_2 \mathbf{j} + c_3 \mathbf{k}.$$

Using the determinant expression in 2(a) we have

$$v \times w = \begin{vmatrix} \mathbf{i} & \mathbf{j} & \mathbf{k} \\ b_1 & b_2 & b_3 \\ c_1 & c_2 & c_3 \end{vmatrix} = (b_2 c_3 - b_3 c_2)\mathbf{i} + (b_3 c_1 - b_1 c_3)\mathbf{j} + (b_1 c_2 - b_2 c_1)\mathbf{k},$$

and therefore

$$u \cdot (v \times w)$$
$$= \left(a_1 \mathbf{i} + a_2 \mathbf{j} + a_3 \mathbf{k}\right) \cdot \left((b_2 c_3 - b_3 c_2)\mathbf{i} + (b_3 c_1 - b_1 c_3)\mathbf{j} + (b_1 c_2 - b_2 c_1)\mathbf{k}\right)$$
$$= a_1(b_2 c_3 - b_3 c_2) + a_2(b_3 c_1 - b_1 c_3) + a_3(b_1 c_2 - b_2 c_1)$$
$$= \begin{vmatrix} a_1 & a_2 & a_3 \\ b_1 & b_2 & b_3 \\ c_1 & c_2 & c_3 \end{vmatrix}.$$

The scalar triple product can be written succinctly in determinant form. Make the important note that the determinant notation *does not* mean that we take absolute values! So this result could be negative or positive. Remember, we are dealing here with vectors and angles.

The scalar triple product and the interpretation of its magnitude as the volume of a parallelepiped is a central feature of multiple integrals in Section 4.H.

**The vector triple product** $u \times (v \times w)$.
As we noted on Page 3, the vector $(v \times w)$ is perpendicular to both $v$ and $w$. Now suppose vector $u$ is not coplanar with $v$ and $w$. What happens if we form the vector cross product of this vector with $(v \times w)$?

Following the same line of reasoning, the result is a new vector which is perpendicular to both $u$ and $(v \times w)$. Given that we have only three dimensions to play with (in $\mathbb{R}^3$), $u \times (v \times w)$ *must lie in the plane defined by the original vectors $v$ and $w$*.

Consequently, $u \times (v \times w)$ must be a linear combination of $v$ and $w$. In fact, by twice applying the determinant formula for the cross product it can easily be verified that

$$u \times (v \times w) = (u \cdot w)v - (u \cdot v)w.$$

Try it!

The subject of this section may seem elementary within the context of straight lines, but the concepts will prove to be quite important when generalized to multivariable scalar and vector function settings in which we deal with *tangent vectors* to more general differential curves.

# 1.B   Introduction to sets

We begin this section with some useful definitions. The reader may refer back to their notes from linear algebra. Alternatively, a good reference is [16].

---

**Definition 1.1**
*Given that $\mathbb{R}^n = \mathbb{R} \times \mathbb{R} \times \cdots \times \mathbb{R}$ is the set of all points $x$ having $n$ independent real components, then $x = (x_1, x_2, \ldots, x_n) \in \mathbb{R}^n$ (where $x_i \in \mathbb{R}$) defines a point in $n$-dimensional Cartesian space.*
*Other notations in common use for a point in $\mathbb{R}^n$ are: $\underset{\sim}{x}$, $\vec{x}$, $\bar{x}$, and $\mathbf{x}$.*

---

■    **Example 1.1:**

The set of points in $\mathbb{R}^2$ is given as $\mathbb{R}^2 = \{(x, y) : x \in \mathbb{R}, y \in \mathbb{R}\}$; a single point with its defining pair of coordinates is shown in Figure 1.8.

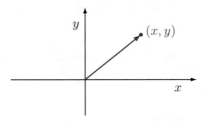

**Figure 1.8**  A point in 2D Cartesian space.

■    **Example 1.2:**

The set of points in $\mathbb{R}^3$ is given as $\mathbb{R}^3 = \{(x, y, z) : x \in \mathbb{R}, y \in \mathbb{R}, z \in \mathbb{R}\}$; a single point with its defining triad of coordinates is shown in Figure 1.9.

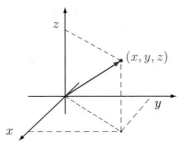

**Figure 1.9**  A point in 3D Cartesian space.

It is not possible to provide a picture of $\boldsymbol{x} \in \mathbb{R}^n$ for $n > 3$ (nor $\mathbb{R}^n$ itself). However, there should be no cause for concern as points in $\mathbb{R}^n$ behave the same as points in $\mathbb{R}^2$ and $\mathbb{R}^3$. That is, they follow the same set of rules. So it is enough to be familiar with points and point operations in $\mathbb{R}^2$ and $\mathbb{R}^3$, and then being able to generalize their properties. The most important point operations are listed below.

**Vector algebra laws.**

Let $\boldsymbol{x}, \boldsymbol{y} \in \mathbb{R}^n$ and $\lambda \in \mathbb{R}$.

At its most basic description, $\mathbb{R}^n$ is an example of a linear vector space which is characterized by the two properties of addition and scalar multiplication:

(a) $\boldsymbol{x} + \boldsymbol{y} = (x_1 + y_1, x_2 + y_2, \ldots, x_n + y_n) \in \mathbb{R}^n$.

(b) $\lambda \boldsymbol{x} = (\lambda x_1, \lambda x_2, \ldots, \lambda x_n) \in \mathbb{R}^n$.

As a direct generalization of the scalar product of 2(c), points in $\mathbb{R}^n$ satisfy the so-called *inner product*,

(c) $\boldsymbol{x} \cdot \boldsymbol{y} = x_1 y_1 + \ldots + x_n y_n \in \mathbb{R}$    — $\mathbb{R}^n$ is called an *inner product space*.

Finally, there are the following generalizations to $\mathbb{R}^n$ of the two fundamental geometric measures:

(d) $|\boldsymbol{x}| = \sqrt{\boldsymbol{x} \cdot \boldsymbol{x}} = \sqrt{x_1^2 + \cdots + x_n^2}$                     — the length of $\boldsymbol{x}$.

(e) $|\boldsymbol{x} - \boldsymbol{y}| = \sqrt{(x_1 - y_1)^2 + \cdots + (x_n - y_n)^2}$
                                                              — the distance between points.

With this distance property $\mathbb{R}^n$ is also a so-called *metric space*, since the distance between points is one measure or metric that allows a geometric characterization of a space.

Using the above definitions one can prove (See Mastery Checks 1.1 and 1.2) some fundamental relations satisfied by position vectors in $\mathbb{R}^n$. These are useful in analysis to establish order relations between vector quantities.

(f) Cauchy-Schwarz inequality:
$$|\boldsymbol{x} \cdot \boldsymbol{y}| \le |\boldsymbol{x}||\boldsymbol{y}|; \quad x_1 y_1 + \ldots + x_n y_n \le \sqrt{x_1^2 + \ldots + x_n^2}\sqrt{y_1^2 + \ldots + y_n^2}.$$

(g) Triangle inequality:    $|\boldsymbol{x} + \boldsymbol{y}| \le |\boldsymbol{x}| + |\boldsymbol{y}|$.

(h) The Cauchy-Schwarz inequality (f) means    $-1 \le \dfrac{\boldsymbol{x} \cdot \boldsymbol{y}}{|\boldsymbol{x}||\boldsymbol{y}|} \le 1$.

Just as in $\mathbb{R}^2$ and $\mathbb{R}^3$, property (h) allows us to define an angle $\theta$ between vectors $\boldsymbol{x}$ and $\boldsymbol{y}$ in $\mathbb{R}^n$ such that

$$\cos \theta = \frac{\boldsymbol{x} \cdot \boldsymbol{y}}{|\boldsymbol{x}||\boldsymbol{y}|}.$$

✍  **Mastery Check 1.1:**

Prove the Cauchy-Schwarz inequality (f).

Hint: Consider the case $n = 2$ before tackling the general theorem. If you square each side of the inequality, you can show that the difference between the results for each side is $\geq 0$. This suggests that a proof might be possible by working backwards.

✍

✍  **Mastery Check 1.2:**

Prove the triangle inequality (g).

Hint: This is achieved easily using (f). Begin by squaring $|\boldsymbol{x} + \boldsymbol{y}|$.

✍

**Points and sets**

In general $\mathbb{R}^n$ we can use property (e), involving the distance between two points, $|\boldsymbol{x} - \boldsymbol{y}|$, to help generalize the "interval" concept to $\mathbb{R}^n$.

In $\mathbb{R}$, the inequality $|x - a| < \epsilon$ (which is equivalent to saying $a - \epsilon < x < a + \epsilon$) describes the set of all $x \in \mathbb{R}$ which lie within $\epsilon$ of $a$. In $\mathbb{R}^n$ we have the analogous case:

---

**Definition 1.2**

*Given $\boldsymbol{a} \in \mathbb{R}^n$, an* **open sphere** *$S_r(\boldsymbol{a}) \subset \mathbb{R}^n$ centred at $\boldsymbol{a}$ and of radius $r$ is the set of all points $\boldsymbol{x} \in \mathbb{R}^n$ that satisfy $|\boldsymbol{x} - \boldsymbol{a}| < r$:*

$$S_r(\boldsymbol{a}) = \{\boldsymbol{x} : \; |\boldsymbol{x} - \boldsymbol{a}| < r\}$$

*for some $r \in \mathbb{R}$. That is, the set of all points $\boldsymbol{x} \in \mathbb{R}^n$ which are no further than $r$ from the given point $\boldsymbol{a}$.*

---

**Remarks**

* The open sphere is non-empty, since it contains $\boldsymbol{a}$ at least.

* In this context $\boldsymbol{a}$ is called the *centre*, and $r$ the *radius* of the set.

The open sphere $S_r(\cdot)$ may now be used to define other point and set properties.

> **Definition 1.3**
> *A point $\boldsymbol{x}$ is called*
>
> - an **interior point** *of a set* $M \subset \mathbb{R}^n$ *if there is an open sphere* $S_r(\boldsymbol{x}) \subset M$ *for some* $r > 0$*;*
>
> - an **exterior point** *of a set* $M \subset \mathbb{R}^n$ *if there is an open sphere* $S_r(\boldsymbol{x}) \not\subset M$ *for some* $r > 0$*;*
>
> - a **boundary point** *of a set* $M \subset \mathbb{R}^n$ *if for any* $r > 0$ *(no matter how small),* $S_r(\boldsymbol{x})$ *contains points in* $M$ *and points not in* $M$*.*

These point definitions are illustrated in Figure 1.10.

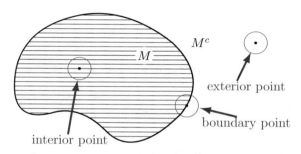

**Figure 1.10**  Interior, exterior, and boundary points to a set.

The reader should work through the following Mastery Checks to consolidate their understanding of these point definitions before going on to set-level concepts.

✎  **Mastery Check 1.3:**

Let $M = \{(x,y) : 0 < x < 1, 0 < y < 1\}$

Draw a diagram of $M$ on the Cartesian plane, showing the points

$P_1(\frac{5}{6}, \frac{5}{6})$, $P_2(1, \frac{1}{3})$, and $P_3(-1, 1)$.

Classify the points as interior, exterior, or boundary.

✎

✎  **Mastery Check 1.4:**

Let $M = \{(x,y) : \dfrac{(x-1)^2}{4} + (y-1)^2 \leq 1\}$.

Draw a diagram of $M$ showing the points $P_1(2,1)$, $P_2(3,1)$, and $P_3(-1,2)$.
Classify the points as interior, exterior, or boundary.

☜

We now establish a framework within which to categorize points that possess
common properties. We begin by grouping points according to Definition 1.3.

---

**Definition 1.4**

- *The set of all interior points of a set $M$ is called the* **interior** *of $M$,
  and denoted* $\mathrm{Int}(M)$:
  $$\mathrm{Int}(M) = \{x : x \in M \text{ and } S_r(x) \subset M \text{ for some } r > 0\}.$$

- *The set of all points not in $M$ is called the* **complement** *of $M$, and
  denoted $M^c$:*
  $$M^c = \{x : x \notin M\}.$$

- *The set of all boundary points of a set $M$ is called the* **boundary** *of
  $M$, and denoted $\partial M$.*

---

It follows from these definitions that $\mathrm{Int}(M^c) \subset M^c$, and if $x$ is an exterior
point to $M$, then $x \in \mathrm{Int}(M^c) \subset M^c$.

The concept introduced in the next definition will play an important role in
our analysis of multivariable functions.

---

**Definition 1.5**
*A set $M$ is called* **open** *if it contains* **only** *interior points.*

---

Accordingly, a set $M$ is an open set if for every point $x \in M$ a positive radius
$r$ can be found such that $S_r(x)$ contains only other points in $M$. Also, only
under the specific condition of set $M$ being open is $\mathrm{Int}(M) = M$. Finally, an
*open neighbourhood* of a point $a \in M$ is an open set $W \in M$.

Although we can utilize the notion of an open set to define a closed set
(see Supplementary problem 7), it proves useful to invoke an independent
concept to define a closed set, that of so-called *limit points*. In this way we

can introduce a notion that is central to our forthcoming discussion on limits.

> **Definition 1.6**
> *A point $a$ of a set $M$ is said to be a* **limit point** *of $M$ if every open sphere $S_r(a)$ contains at least one point of $M$ different from $a$.*

This means that there are points in $M$ that are arbitrarily close to $a$. Hence, in approaching $a$ from within $M$ we are always guaranteed to encounter other points in $M$.

Introducing limit points not only allows for a meaningful definition of a closed set, but it also allows one to readily prove a number of facts about closed sets, some of which are included as exercises at the end of this chapter. First the definition:

> **Definition 1.7**
> *A set $M$ of $\mathbb{R}^n$ is said to be* **closed** *if it contains all its limit points.*

And, intimately related to this definition is the concept of set *closure*. For our purposes we invoke the following definition.

> **Definition 1.8**
> *The* **closure** *of a set $M$, denoted $\overline{M}$, is the union of $Int(M)$ with its boundary:*
> $$\overline{M} = Int(M) \cup \partial M = \{x : x \in Int(M) \text{ or } x \in \partial M\}.$$

Alternatively, $\overline{M}$ can be defined as the union of $M$ and the set of all its limit points, $L_M$. It can be shown (see Supplementary problem 7) from this definition that (a) a closed set $M$ is equal to its closure $\overline{M}$, and (b) that a set is closed if and only if it contains its boundary. Along this same line of thought, an alternative consequence of Definitions 1.5–1.8 is that the boundary of a set $M$ contains those points that are in common with the closure of a set $M$ and the closure of its complement, $M^c$. In other words

$$\partial M = \{x : x \in \overline{M} \cap \overline{M^c}\}.$$

The concept of set *boundedness* arises in both contexts of differential and integral calculus of multivariable functions. As the term suggests it essentially relates to a set being limited in geometric extent.

> **Definition 1.9**
> A set $M \subset \mathbb{R}^n$ is called **bounded** *if there exists a $K \in \mathbb{R}$ such that $|\boldsymbol{x}| < K$ for all $\boldsymbol{x} \in M$.*

Definition 1.5 is utilized in the definition of derivatives in Chapter 2, while Definitions 1.4–1.9 are invoked in Chapter 3 and 4, although they are also used implicitly elsewhere.

✐ **Mastery Check 1.5:**

(1) Let $M = \{(x,y) : 0 < x < 1, 0 < y < 1\}$. What is the set $\partial M$?

(2) Let $M = \left\{(x,y) : \dfrac{(x-1)^2}{4} + (y-1)^2 \leq 1\right\}$. What is the set $\partial M$?

✐

The next definition is most useful when invoked together with function continuity to establish conditions that guarantee certain function behaviour. We shall see this employed in practice in Sections 3.B, 4.A and 4.D, but also 1.C.

> **Definition 1.10**
> A set $M \subset \mathbb{R}^n$ is called **compact** *if it is both* **closed** *and* **bounded**.

✐ **Mastery Check 1.6:**

For each of the sets $M$ given below, answer the following questions:

Is $M$ bounded? If it is, find a number $K$ such that $|\boldsymbol{x}| < K$ for all $\boldsymbol{x} \in M$?

Is $M$ compact? If it is not, write down the closure $\overline{M} = M \cup \partial M$.

Then draw a diagram showing $M$, $\partial M$, and $K$.

(1) Let $M = \{(x,y) : 0 < x < 1, 0 < y < 1\}$.

(2) Let $M = \left\{(x,y) : \dfrac{(x-1)^2}{4} + (y-1)^2 \leq 1\right\}$.

✐

## 1.C   Real-valued functions

**Basic concepts and definitions.**

In Chapters 2, 3, and 4, we focus attention almost exclusively on scalar-valued functions of many variables, while in Chapter 5 we extend the ideas to vector-valued functions. In both contexts the following introduction to fundamental properties of multi-valued functions is invaluable. To start, we introduce some more notation and a pictorial view of what functions do.

In single-variable calculus we have the following scenario:

Let $y = f(x)$. The "graph" of $f$ is the set of ordered pairs $\{(x, f(x))\} \in \mathbb{R}^2$. This is shown graphically in Figure 1.11 where the independent variable $x$ and dependent variable $y$ are plotted on mutually orthogonal axes.

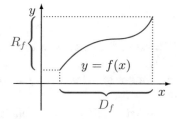

**Figure 1.11**   The Cartesian graph of $y = f(x)$.

This way of visualizing functions of one variable was introduced in the early 17<sup>th</sup> century by René Descartes [17], and is named the Cartesian representation in recognition. It is quite a useful means of illustrating function dependence and function properties, especially for functions of one or two variables.

It ceases to be as useful, however, for functions of more than two variables. For the latter cases one resorts to simply considering a set-mapping picture. For the case $y = f(x)$ this is a simple interval-to-interval map as shown in Figure 1.12.

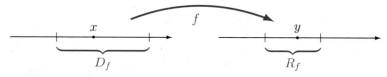

**Figure 1.12**   The set map of $D_f \subset \mathbb{R} \longrightarrow R_f \subset \mathbb{R}$.

For real-valued functions of many variables: $y = f(\boldsymbol{x}) = f(x_1, x_2, \ldots, x_n)$, the corresponding illustrative representation is shown in Figure 1.13. The left-hand $x$-interval in the single-variable calculus case is replaced by a more general $\boldsymbol{x}$-region for the multivariable case.

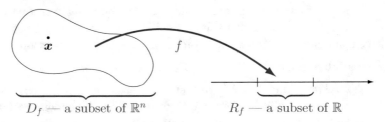

$$D_f \text{ — a subset of } \mathbb{R}^n \qquad\qquad R_f \text{ — a subset of } \mathbb{R}$$

**Figure 1.13**  The set map of $D_f \subset \mathbb{R}^n \longrightarrow R_f \subset \mathbb{R}$.

---

**Definition 1.11**

*Consider a real-valued function $f$ of one or more variables, whose graph is the point set $\{(\boldsymbol{x}, f(\boldsymbol{x}))\}$.*

*The **domain** $D_f$ of $f$ is the set of all values of $\boldsymbol{x}$ for which $f$ is defined (that is, for which $f$ makes sense).*

*The **range** $R_f$ of $f$ is the set of all possible values of $y = f(\boldsymbol{x})$ for all $\boldsymbol{x} \in D_f$.*

---

In our multivariable setting $\boldsymbol{x} \in \mathbb{R}^n$ is the independent multivariable and $y \in \mathbb{R}$ is (still) the dependent variable. It is common to find the following terminology used in text books. The independent variable, here $\boldsymbol{x} \in \mathbb{R}^n$, is sometimes referred to as the *pre-image* of $y$, while the dependent variable, here $y \in \mathbb{R}$, is called the *image* under $f$. The function $f$ is alternatively referred to as a *mapping* from one set to another, or an *operation* that takes a point, here $\boldsymbol{x}$, to $y$, or a *rule* that associates $y$ with a point $\boldsymbol{x}$. As far as mathematical notation is concerned, the mapping under $f$ is either described *pointwise*

$$f : \boldsymbol{x} \longmapsto y$$

or in a *set-wise* sense:

$$f : D_f \subseteq \mathbb{R}^n \longrightarrow R_f \subseteq \mathbb{R}.$$

Both references will be used in this book.

### ■  Example 1.3:

Consider $f(x, y) = \sqrt[3]{y - x^2}$.

Now, recall from single-variable calculus that $s = t^{1/3}$ is defined for all $t$, since

$$\text{if } t > 0, \text{ then } s > 0;$$
$$\text{if } t < 0, \text{ then } s < 0;$$
$$\text{if } t = 0 \text{ then } s = 0.$$

In addition, we readily see that $y - x^2$ can take any real value. Combining these facts we deduce that $f$ is defined everywhere. This implies that $D_f = \mathbb{R}^2$ and $R_f = \mathbb{R}$.

■

### ✍  Mastery Check 1.7:

Find the (implied) domain and range of the following functions:

1. $f(x) = \sqrt{16 - x^4}$;
2. $f(x) = \dfrac{3 - x^2}{1 + x^2}$;
3. $\mathrm{gd}(x) = \sin^{-1}\left(\tanh(x)\right)$;

4. $f(x, y) = \sqrt{9 - x^2 - y^2}$;
5. $f(x, y) = \ln(x - y)$;
6. $f(x, y, z) = \ln(|z - x^2 - y^2|)$.

(The function $\mathrm{gd}(x)$ is known as the Gudermannian function.)

✍

### ■  Example 1.4:

Suppose $f(x, y) = \sin^{-1}(x^2 + y^2)$.

Before considering this multivariable function, recall from single-variable calculus that within the intervals $-\dfrac{\pi}{2} \le z \le \dfrac{\pi}{2}$ and $-1 \le w \le 1$,

$$z = \sin^{-1} w \iff w = \sin z.$$

The graphs of these *inverse* functions are shown in Figure 1.14.

Note that in our case $w = x^2 + y^2 \ge 0$, and therefore so is $z \ge 0$. So,

$$|w| \le 1 \implies |x^2 + y^2| \le 1 \implies 0 \le x^2 + y^2 \le 1.$$

This defines the unit disc in the $xy$-plane. That is, $D_f$ is the unit disc (the unit circle and its interior).

Similarly,   $|z| \le \dfrac{\pi}{2} \implies 0 \le z \le \dfrac{\pi}{2}$ since $z \ge 0$. So, $R_f = \left[0, \dfrac{\pi}{2}\right]$.

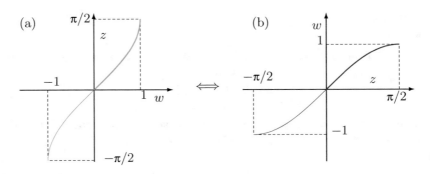

**Figure 1.14**  Graphs of the functions (a) $z = \sin^{-1} w$ and (b) $w = \sin z$.

■

✍  **Mastery Check 1.8:**

Consider the function $f(x, y) = \ln\left(2 - e^{x^2 + y^2}\right)$. Find the implied domain and range.

✍

Although we will pay considerable attention to real-valued functions of several variables, we shall see in Chapter 5 that there is another important class of functions of several variables. These are vector-valued multivariable functions. Examples include:

(a) $\boldsymbol{f} : t \longmapsto \boldsymbol{y} \in \mathbb{R}^p$

— vector-valued functions of one real variable, $t$.

$$\boldsymbol{f}(t) = (f_1(t), f_2(t), \ldots, f_p(t))$$

(b) $\boldsymbol{f} : \boldsymbol{x} \in \mathbb{R}^n \longmapsto \boldsymbol{y} \in \mathbb{R}^m$

— vector-valued functions of a vector variable (several real variables).

$$\boldsymbol{f}(\boldsymbol{x}) = (f_1(\boldsymbol{x}), f_2(\boldsymbol{x}), \ldots, f_m(\boldsymbol{x}))$$

**Limits and continuity.**

In the next chapter we introduce and explore the concept of partial differentiation. In the lead up to that discussion it will be necessary to explain a number of concepts we shall then take for granted. Most importantly there is the notion of *function continuity*. For multivariable functions this will be discussed in detail in Section 2.B, but we can set the stage here with a short review of the subject as it relates to functions of one variable.

Function continuity is defined in terms of limiting processes. Mention has already been made of limit points of closed sets. We said that a point $a$ is a limit point if any open sphere centred on $a$, no matter how small in radius, contains points other than $a$.

Similarly, segments of the real line possess the property that any open *interval* $I$, no matter how small, centred on a point $a$, contain points $x$ in $I$ different from $a$. The real line and any of its finite segments are therefore said to be *complete*: containing no gaps. This conjures up the notion of a set *continuum*, moving smoothly from one real value to another, never meeting any holes.

This notion gives critical meaning to the formalism $x \to a$ as the process of approaching a real value $a$ along the real line. To be even more precise, we specify $x \to a^-$ and $x \to a^+$ as meaning the respective approaches to $a$ along the real line from "below" $a$ $(x < a)$ and from "above" $a$ $(x > a)$.

Now with thought given to single-variable functions defined on a domain $D_f \subset \mathbb{R}$, the different approaches $x \to a^-$ and $x \to a^+$ for $a, x \in D_f$ can have all manner of implications for the function. Assuming $a, x \in D_f$ we define the *process* of taking a limit of a function, which we denote either by

$$\lim_{x \to a^-} f(x),\ \lim_{x \to a^+} f(x),\ \text{or}\ \lim_{x \to a} f(x)$$

as considering the sequence of values $f$ progressively takes as $x \to a^-$, $x \to a^+$, or in their combination. These considerations are of course separate to the question of what value $f$ actually takes at $a$. To summarize all of these ideas we have the following definition.

---

**Definition 1.12**

*A function is said to be* **continuous** *at an interior point $a$ of its domain $D_f \subset \mathbb{R}$ if*

$$\lim_{x \to a} f(x) = f(a).$$

*If either the equality is not satisfied, or the limit fails to exist, then $f$ is said to be discontinuous at $a$.*

---

To reiterate, in the context of functions of a single variable the above limit is equivalent to the requirement that the limits approaching $a$ from below $(x < a)$ and from above $(x > a)$ exist and are equal. That is, the single expression $\lim_{x \to a} f(x)$ means that

$$\lim_{x \to a^-} f(x) \quad = \quad \lim_{x \to a^+} f(x) \quad = \quad A.$$

$$\left\{ \begin{array}{c} \text{left-hand-side} \\ \text{limit of } f(x) \end{array} \right\} = \left\{ \begin{array}{c} \text{right-hand-side} \\ \text{limit of } f(x) \end{array} \right\}$$

If the above equality is not satisfied we say that the limit does not exist. Definition 1.12 then also stipulates that for continuity the common limiting value, the aforementioned $A$, must also equal the value of the function $f$ at $x = a$, $f(a)$.

The following example demonstrates graphically some different circumstances under which a limit of a function exists or does not exist, and how these relate to the left and right limits. Note the sole case of the function value actually being specified (solid dot) in the left-most graph in Figure 1.15. Is either function continuous at $a$?

■   **Example 1.5:**

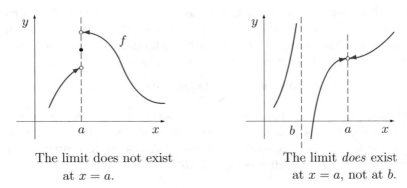

The limit does not exist
at $x = a$.

The limit *does* exist
at $x = a$, not at $b$.

**Figure 1.15**  When does a limit exist?

■

From the definition and subsequent discussion we are led to an important theorem of pointwise continuity.

---

**Theorem 1.1**

*A function $f$ is continuous at an interior point $a$ if and only if*

$$\lim_{x \to a^-} f(x) = \lim_{x \to a^+} f(x) = f(a).$$

In slightly more practical mathematical language the statement of Definition
1.12 and Theorem 1.1 can be expressed by the following:

$$0 < |x - a| < \delta, \ x, a \in D_f \quad \Longrightarrow \quad |f(x) - A| < \epsilon \text{ for some } \delta = \delta(\epsilon).$$

Graphically, this limit definition can be represented as in Figure 1.16 below.

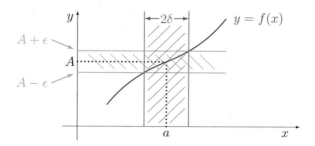

**Figure 1.16**   The $\epsilon$–$\delta$ condition.

The concepts developed above will be employed in Section 2.B. For now, this
pointwise concept can be extended to the entire function domain.

**Definition 1.13**
*A function $f$ that is continuous at every point of its domain $D_f \subset \mathbb{R}$ is
said to be* **continuous** *over that domain.*

We denote by $C(D_f)$ the set of all continuous functions defined on the domain
$D_f$.

Still on the single-variable case, we will have need in Sections 3.C, 4.A and
4.B of the following important theorem, which combines the concepts of function continuity and domain compactness to give an important result.

**Theorem 1.2**
*Let $f : \mathbb{R} \longrightarrow \mathbb{R}$ be continuous on a closed and bounded interval $D_f \subset \mathbb{R}$.
Then $f$ attains an absolute maximum and an absolute minimum value in
$D_f$. That is, there exist values $x_L \in D_f$ and $x_U \in D_f$ such that*

$$f(x_L) \leq f(x) \leq f(x_U) \text{ for all } x \in D_f.$$

A moment's thought and possibly some simple sketches will make this theorem intuitively obvious. The self-evidence of the theorem, however, does not diminish its significance.

We end this section with a short catalogue of well-established results that can assist us in evaluating limit processes for both single-variable and multi-variable functions, where the latter cases comprise single-variable functions. Three of the squeeze relations listed below are featured in Figure 1.17.

**Some useful standard limits:**

$* \displaystyle\lim_{x \to 0} \frac{\sin x}{x} = 1.$ $\qquad\qquad * \displaystyle\lim_{x \to \infty} \frac{\ln x}{x^\alpha} = 0$ for constant $\alpha > 0.$

$* \displaystyle\lim_{x \to 0} \frac{\tan x}{x} = 1.$ $\qquad\qquad * \displaystyle\lim_{x \to 0} x^\alpha \ln x = 0$ for constant $\alpha > 0.$

$* \displaystyle\lim_{t \to 0} \frac{\sin(xt)}{t} = x.$ $\qquad\qquad * \displaystyle\lim_{t \to 0} \frac{\cos t - 1}{t} = 0.$

**Some useful squeeze relations:**

(a) $\sin x < x < \tan x, \quad 0 < x < \dfrac{\pi}{2}.$

(b) $x < e^x - 1 < \dfrac{x}{1 - x}, \quad 0 < x < 1.$

(c) $\dfrac{x}{1 + x} < \ln(1 + x) < x,$ for $x > -1, \neq 0.$

(d) $e^x > 1 + x \quad \forall x \neq 0.$

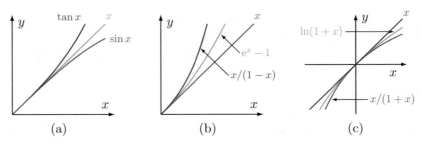

**Figure 1.17** Graphs of various squeeze relations relative to $y = x.$

## 1.D   Coordinate systems

Up until now we have represented points in $\mathbb{R}^2$ and $\mathbb{R}^3$ in terms of *Cartesian* coordinates, $(x, y)$ as in Example 1.1 and $(x, y, z)$ as in Example 1.2. However, problems arise that are better described in other coordinate systems. Such problems arise in both the differential and integral calculus (Sections 3.E, 4.E, and 4.H) and are usually associated with the geometry of the region under consideration. The most common coordinate systems that we will encounter are the *polar coordinate* system in $\mathbb{R}^2$, and the *cylindrical* and *spherical coordinate* systems in $\mathbb{R}^3$. Note that there are other standard systems that can be useful in specific cases (see [15]) and even non-standard systems may be needed to solve some problems (see Section 4.E).

There are three general features to note. First, the 2D Cartesian and polar coordinate systems have the same origin. Similarly, the 3D Cartesian and cylindrical or spherical coordinate systems have a common origin. Second, the non-Cartesian coordinates are designed to uniquely identify and represent every point in $\mathbb{R}^2$ or $\mathbb{R}^3$, as do their Cartesian counterparts. That is, these coordinate systems span the whole of $\mathbb{R}^2$ and $\mathbb{R}^3$, respectively. Finally, the individual coordinate variables within a given non-Cartesian system are independent of each other, just as the individual Cartesian coordinates are independent variables in the Cartesian system.

### Polar coordinates

Consider an arbitrary point $P$ in the plane with Cartesian coordinates $(x, y)$. $P$'s distance from the origin is

$$r = \sqrt{x^2 + y^2},$$

while the angle between $P$'s position vector $\boldsymbol{r} = (x, y)$ and the $x$-axis is given by

$$\tan \theta = \frac{y}{x}.$$

The unique inverse relation is given by the pair of equations

$$\left. \begin{array}{l} x = r\cos\theta \\ y = r\sin\theta \end{array} \right\} \quad \text{for} \quad 0 \le \theta \le 2\pi.$$

Thus, every point in $\mathbb{R}^2$ can be uniquely represented by the pair of so-called *polar* coordinates $(r, \theta)$ defined on the domain $[0, \infty) \times [0, 2\pi]$. The relationship between the two coordinate representations is shown in Figure 1.18(a).

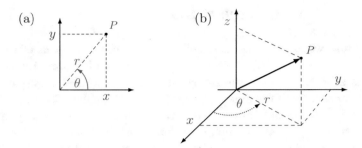

**Figure 1.18** (a) 2D polar and (b) 3D cylindrical coordinates.

The distance $D$ between two arbitrary points $P(x_1, y_1)$ and $Q(x_2, y_2)$ is then

$$D = \left[ (x_1 - x_2)^2 + (y_1 - y_2)^2 \right]^{1/2}$$
$$= \left[ r_1^2 + r_2^2 - 2r_1 r_2 \cos(\theta_1 - \theta_2) \right]^{1/2},$$

where $x_i = r_i \cos \theta_i$ and $y_i = r_i \sin \theta_i$, $i = 1, 2$.

### Cylindrical polar coordinates

An arbitrary point $P$ in 3D is defined by Cartesian coordinates $(x, y, z)$. The preceding case of plane polar coordinates is thus easily generalized to *cylindrical polar coordinates* in 3D by the inclusion of the Cartesian coordinate $z$ to account for the third dimension.

We therefore have the relations

$$x = r \cos \theta, \ y = r \sin \theta, \ z = z.$$

Figure 1.18(b) shows the point $P$ represented by the two alternative coordinate systems $(x, y, z)$ and $(r, \theta, z)$. The distance between any two points $P$ and $Q$ generalizes to

$$D = \left[ (x_x - x_2)^2 + (y_1 - y_2)^2 + (z_1 - z_2)^2 \right]^{1/2}$$
$$= \left[ r_1^2 + r_2^2 - 2r_1 r_2 \cos(\theta_1 - \theta_2) + (z_1 - z_2)^2 \right]^{1/2}.$$

### Spherical polar coordinates

The second generalization to 3D of polar coordinates is the *spherical polar* coordinate system. This is based on the notion of defining a point on a sphere in terms of latitude and longitude angles. To be precise, an arbitrary point $P$ in 3D with Cartesian coordinates $(x, y, z)$ is identified by the triplet of independent variables $(\rho, \phi, \theta)$ defined by

$$\left.\begin{array}{l} x = \rho \sin\phi \cos\theta, \\ y = \rho \sin\phi \sin\theta, \\ z = \rho \cos\phi, \end{array}\right\} \qquad \begin{array}{l} 0 \le \rho < \infty, \\ 0 \le \phi \le \pi, \\ 0 \le \theta \le 2\pi, \end{array}$$

with the inverse relations

$$\rho^2 \;=\; x^2 + y^2 + z^2 \;=\; r^2 + z^2, \quad \tan\theta \;=\; \frac{y}{x},$$

$$\cos\phi \;=\; \frac{z}{(r^2 + z^2)^{1/2}} \quad \text{or} \quad \sin\phi = \frac{r}{\rho}.$$

Figure 1.19 illustrates how the variables are related geometrically. The origins of the angles $\phi$ and $\theta$ as $z$−axis and $x$−axis, respectively, are also indicated.

The distance between two arbitrary points $P$ and $Q$ in 3D is now expressed

$$D^2 = \rho_1^2 + \rho_2^2 - 2\rho_1\rho_2 \cos(\phi_1 - \phi_2) - 2\rho_1\rho_2 \sin\phi_1 \sin\phi_2 \big( \cos(\theta_1 - \theta_2) - 1 \big).$$

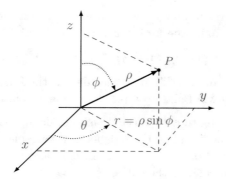

**Figure 1.19**   3D spherical coordinates.

# 1.E   Drawing or visualizing surfaces in $\mathbb{R}^3$

Throughout the book and indeed throughout the subject generally we will need to recognize, but also to sketch or otherwise visualize, areas and volumes embedded in $\mathbb{R}^2$ and $\mathbb{R}^3$, respectively. The ability to conceptualize regions in 2D and 3D makes the task of understanding multivariable function behaviour easier. Also, in the case of integration in Chapter 4, it simplifies the task of establishing integration limits that define the boundaries of a region. Lastly, being able to visualize structures in 3D can be helpful when checking the reasonableness of possible solutions to mathematical exercises.

Most of the examples and exercises to follow utilize MATLAB® in the visualization of various surfaces (using the "`surf`" function). However, the reader with access to other graphing software should be able to translate the programming information shown below into relevant code for their own graphing tool.

As an alternative the first example that follows shows how to piece together a picture of a graph with little or no reliance on software. A more detailed discussion of this manual approach (restricted to functions of a single variable) can be found in Chapter 5 of [1].

■   **Example 1.6:**

We shall consider here the graph of the function

$$f(x, y) = \frac{4x}{1 + x^2 + y^2} \quad \text{for} \quad (x, y) \in \mathbb{R}^2.$$

This function features in an exercise in a later chapter. For now we are just interested in determining the form taken by the function's graph,

$$G = \left\{ (x, y, z) : (x, y) \in \mathbb{R}^2, z = f(x, y) \right\}.$$

In the steps that follow we will in essence dissect the function, and with the pieces we obtain we will build up a picture of the graph.

**Step 1:** The first thing to note is the domain of definition. What you would be looking for are the limits on the independent variables as well as possible points where the function is not defined. In our case, the function is defined everywhere so the domain is the entire $xy$-plane.

**Step 2:** The second thing to do is to look for any zeros of the function. That is, we look for intercept points in the domain at which the function takes the value zero. Here, $f = 0$ when $x = 0$, that is, at all points along the $y$-axis.

**Step 3:** We now look for any symmetry. We note that the function is odd in $x$ but even in $y$. The symmetry in $x$ means that for any fixed $y$ — which means taking a cross-section of the graph parallel to the $x$-axis — howsoever the graph appears for $x > 0$, it will be inverted in the $xy$-plane for $x < 0$.

The symmetry in $y$ means that for any fixed $x$ (that is, a cross-section parallel to the $y-$axis) the graph will look the same on the left of $y = 0$ as on the right. Note, however, that because of the oddness in $x$, the graph will sit *above* the $xy$-plane for $x > 0$, but *below* the plane for $x < 0$.

So far, we have the impressions shown in Figure 1.20.

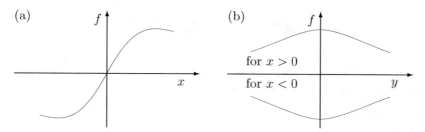

**Figure 1.20**   The function $f$ is odd in $x$ (a), but even in $y$ (b).

**Step 4:** It is often instructive to look at small values of $x$ and $y$.

Supposing $|x|$ to be very small compared to both 1 and $y$ we see that $f$ behaves as

$$f \approx \left(\frac{4}{1+y^2}\right)x \quad \text{as} \quad |x| \longrightarrow 0.$$

(The symbol $\approx$ means "approximately equal to" and indicates very close correspondence.) We see that $f$ behaves linearly with respect to $x$, with a coefficient that depends on $y$. This is consistent with the fact that $f$ is odd in $x$ (Figure 1.20(a)).

Next, supposing $|y|$ to be very small compared to 1 and $x$, the function will tend to

$$f \approx \frac{4x}{1+x^2} \quad \text{as} \quad |y| \longrightarrow 0.$$

That is, $f$ behaves very much as a constant with respect to $y$, dependent only on the given value of $x$. Again this is consistent with our finding that $f$ is even in $y$: we would expect the function to be approximately constant for small $|y|$, as in Figure 1.20(b).

**Step 5:** We now consider the behaviour of $f$ for large values of $x$ and $y$, the "asymptotic" behaviour of $f$.

Again fixing $y$ and taking $|x|$ very large compared with either 1 or $y$, the function will tend to behave as

$$f \approx \frac{4}{x} \longrightarrow 0 \quad \text{as} \quad x \longrightarrow \pm\infty.$$

The approach to zero will depend on the sign of $x$: approaching zero from *above* for $x > 0$ and from *below* for $x < 0$. (See Figure 1.21(a).)

On the other hand, fixing $x$ instead and taking $|y|$ very large compared to either 1 or $x$, we find that

$$f \approx \frac{4x}{y^2} \longrightarrow 0 \quad \text{as} \quad y \longrightarrow \pm\infty.$$

So the function again approaches zero. Note again the sign difference for positive and negative $x$ (Figure 1.21(b)).

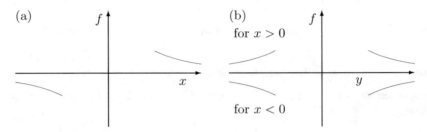

**Figure 1.21** The behaviour of $f$ for large $|x|$ (a), and large $|y|$ (b).

Now let's see if we can deduce something from this incomplete picture.

The function *is* zero along $x = 0$, and it *tends* to zero at large $|x|$ (and large $|y|$) and is nonzero in between.

Since the function does not have any singular behaviour anywhere in $\mathbb{R}^2$, we can conclude that there must be at least one point along $x > 0$ where the function peaks at some positive value, and similarly there must be at least one point along $x < 0$ where the function bottoms at some negative value. We are thus led to ...

**Step 6:** Investigate $f$ for maxima and minima. This step we will leave until we have at our disposal the differentiation tools developed in the next chapter and applied in Chapter 3 (see Mastery Check 3.7). But the above information is enough to put together the sketch shown in our final Figure 1.22.

If we look at the curves running parallel to the $x$−axis we see that the extremities (large $|x|$) match the predictions in Figure 1.21(a), while the middle sections agree with the curves in Figure 1.20(a). Regarding the curves parallel to the $y$-axis, the extremes (large $|y|$) concur with Figure 1.21(b), while the sections crossing the $x$-axis agree with the lines in Figure 1.20(b).

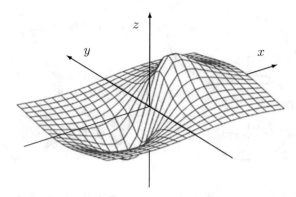

**Figure 1.22**  Putting it all together.

■

This next example similarly shows how one can visualize a surface without using graphing software.

■   **Example 1.7:**

Consider $S = \{(x, y, z) : x^2 + y^2 + z^2 = a^2, \ a > 0\}$. This is a surface in $\mathbb{R}^3$, Figure 1.23; it is a surface because there exists a relation between the three variables $(x, y, z)$. They are no longer completely independent: one variable can be considered a function of the other two.

Now set $z = 0$. This simplifies to the subset satisfying $x^2 + y^2 = a^2$ which is a curve (circle) in the $xy$-plane. Note that these two equations for the three variables, which is equivalent to setting two conditions on the three variables, generate a curve in $\mathbb{R}^2$.

A consistent interpretation is that of the intersection of two surfaces: The plane $z = 0$ and the sphere $S$ giving rise to the subset of points the surfaces have in common — the circle of radius $a$ in the $xy$-plane.

Suppose that $a > 2$, say, in $S$. Then setting

$$z = 0 \implies x^2 + y^2 = a^2,$$
$$z = 1 \implies x^2 + y^2 = a^2 - 1 < a^2,$$
$$z = 2 \implies x^2 + y^2 = a^2 - 4 < a^2 - 1 < a^2,$$
$$z = a \implies x^2 + y^2 = a^2 - a^2 = 0 \iff x = y = 0.$$

These are examples of *level sets* defining circles in the $xy$-plane. We will come back to discuss these in detail in Section 1.F.

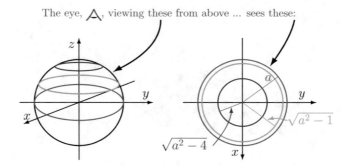

The eye, ⋀, viewing these from above ... sees these:

**Figure 1.23**  The sphere of radius $a$ and a few of its level sets.

### ■  Example 1.8:

The same example as Example 1.7, but now using  MATLAB®: Figure 1.24.

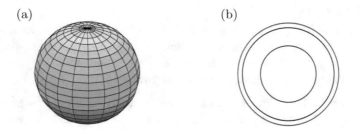

(a)                                                      (b)

**Figure 1.24**  The sphere of radius 2.4 (a) and its level sets (b).

This version of the  MATLAB® code produces figures without axes.

```
figure(1)
[X,Y,Z]=sphere; % generates three 21X21 matrices for a unit
sphere
X=2.4*X; Y=2.4*Y; Z=2.4*Z; % sphere now has radius 2.4
surf(X,Y,Z), axis tight, axis off
figure(2)
[X,Y,Z]=sphere(100); % better definition with 100 segments
X=2.4*X; Y=2.4*Y; Z=2.4*Z;
map=[1,0,0;0,0.2,0;0,0,1]; % colours are red, dark green, blue
contour(X,Y,Z,[0,1,2]), axis off
colormap(map);
```

### ✎  Mastery Check 1.9:

Set up your own matrices in MATLAB® for the `surf` plot, as follows (or otherwise), and draw the sphere again:

```
r=2.4; theta=linspace(0,2*pi,21); phi=linspace(0,pi,21);
X=r*sin(phi')*cos(theta);
Y=r*sin(phi')*sin(theta);
Z=r*cos(phi')*ones(1,21);
surf(X,Y,Z)
```

✎

### ■  Example 1.9:

A circular cone: $S = \{(x,y,z) : z^2 = x^2 + y^2, -1 \leq x, y \leq 1\}$

Let $0 \leq \theta \leq 2\pi$ and $0 \leq r \leq 1$, $\left. \begin{array}{l} x = r\cos\theta \\ y = r\sin\theta \end{array} \right\} \implies z = \pm r.$

This example illustrates why care should be exercised in cases involving squares. It is easy to forget that there is some ambiguity when taking the square root: see Figure 1.25.

The MATLAB® default figure format has tick marks with labels on the axes which suit most purposes, and there are simple functions for producing labels for the axes themselves, as shown in the sample code that follows.

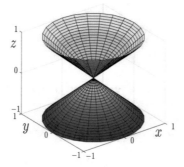

**Figure 1.25**  The graph of $z = \pm\sqrt{x^2 + y^2}$.

The MATLAB® code:

```
theta=linspace(0,2*pi,25);
r=linspace(0,1,25); % 25 intervals from 0 to 1
x=cos(theta')*r; y=sin(theta')*r; Z=sqrt(x.^2+y.^2);
```

```
surf(x,y,Z), hold on, surf(x,y,-Z)
xlabel('x'), ylabel('y'), zlabel('z')
```

However, the following may be used in place of the last line of code to produce clearer labels, given here for the $x$-axis, and easily adapted for the other two axes.

```
ax=gca; % Get the axis handle, call it 'ax'
xticks([-1,0,1]); % This sets the tick marks
% Place the new larger axis label at (0.6,-1.2,-1)
text(0.6,-1.2,-1,'$x$','interpreter','latex','fontsize',24)
% Overwrite the tick labels with blanks
ax.XAxis.TickLabels=' ',' ',' ';
% Place the new x-tick label '-1' at (-1,-1.2,-1)
text(-1,-1.2,-1,'-1','fontsize',16)
text(0,-1.2,-1,'0','fontsize',16)
text(1,-1.2,-1,'1','fontsize',16)
```

        ■

### ✍ Mastery Check 1.10:

Consider these conic sections for the case in Example 1.9:

1. Set $y = 0 \implies z^2 = x^2 \implies z = \pm x$ — a pair of straight lines.

2. Set $y = 0.5 \implies z^2 = x^2 + 0.25 \implies (z - x)(z + x) = 0.5^2$
   — a hyperbola.

3. Set $z = 0.6 \implies x^2 + y^2 = 0.6^2$ — a circle, radius 0.6.

4. Set $z = y + 0.25 \implies x^2 + y^2 = y^2 + 0.5y + 0.0625$
   $\implies y = 2x^2 - 0.125$ — a parabola.

Each of these curves may be generated using MATLAB®.

Your task is to add in turn the groupings of lines of the following code to the end of the code for the cone, then use the "Rotate 3D" button on the MATLAB® figure to view the curves in space.

```
% the line pair
title('$z^2=x^2+y^2$','interpreter','latex')
x=linspace(-1,1,11)'; y=zeros(11,1); Z=x*ones(1,11);
surf(x,y,Z)
% the hyperbola
title('$z^2=x^2+y^2$','interpreter','latex')
```

```
x=linspace(-1,1,11)'; y=0.5*ones(11,1); Z=x*ones(1,11);
surf(x,y,Z)
% the circle
title('$z^2=x^2+y^2$','interpreter','latex')
x=linspace(-1,1,11); y=x; Z=0.6*ones(11);
surf(x,y,Z)
% the parabola
title('$z^2=x^2+y^2$','interpreter','latex')
x=linspace(-1,1,11)'; y=x; Z=(y+0.25)*ones(1,11);
surf(x,y,Z)
```

Figures 1.26(a) and (b) are for the last one of these, at two different aspects (see if you can get these plots):

`view(0,90)`, and `view(-90,4)`

Conic sections: The parabola.

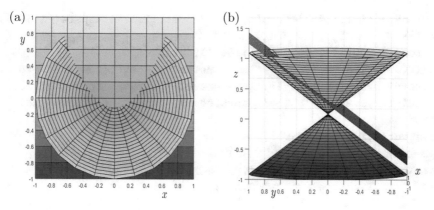

**Figure 1.26** Two views of the intersection of the graphs of $z^2 = x^2 + y^2$ and $y = x$: (a) top view, (b) side view.

■   **Example 1.10:**

The hyperbolic paraboloid of Figure 1.27: $S = \{(x, y, z) : z = 1 + x^2 - y^2\}$.

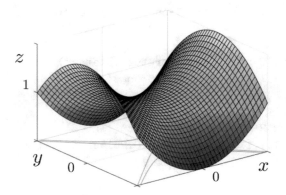

**Figure 1.27** The graph of $z = 1 + x^2 - y^2$ with two level sets.

Setting $z =$ constant will give a hyperbola. For example, set $z = 0.9$, $y^2 - x^2 = 0.1$, or set $z = 0.99$, $y^2 - x^2 = 0.01$.

These curves shown in the figure are projections onto the $xy$-plane ($z = 0$). The MATLAB® code:

```
X=linspace(-1,1,41)'*ones(1,41); Y=ones(41,1)*linspace(-1,1,41);
Z=1+X.^2-Y.^2 ; surf(X,Y,Z)
xticks([-1,0,1]), yticks([-1,0,1]), zticks([0,1,2])
text(0.7,-1.2,-0.1,'$x$','interpreter','latex','fontsize',28)
text(-1.3,0.7,-0.1,'$y$','interpreter','latex','fontsize',28)
text(-1.2,1.2,1.7,'$z$','interpreter','latex','fontsize',28)
hold on
contour(X,Y,Z,[0.9,0.99])
ax=gca;
ax.XAxis.TickLabels=' ',' ',' ';
ax.YAxis.TickLabels=' ',' ',' ';
ax.ZAxis.TickLabels=' ',' ',' ';
text(0,-1.2,-0.1,'$0$','interpreter','latex','fontsize',20)
text(-1.3,0,0,'$0$','interpreter','latex','fontsize',20)
text(-1,1.3,1,'$1$','interpreter','latex','fontsize',20)
```

∎

## ✍ Mastery Check 1.11:

The hyperbolic paraboloid: $S = \{(x, y, z) : z = 1 + x^2 - y^2\}$.

Setting $y = 0$ will give the parabola $z = 1 + x^2$. Setting $x = 0$ will give the parabola $z = 1 - y^2$.

Produce 3D plots for each of these.

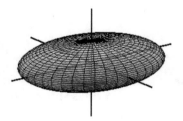

## ■ Example 1.11:

**How to draw an ellipsoid in** MATLAB®.

The following code produces the graph of the ellipsoid $\dfrac{x^2}{4^2} + \dfrac{y^2}{3^2} + \dfrac{z^2}{2^2} = 1$ shown in Figure 1.28, whose semi-axes are $a = 4$, $b = 3$, and $c = 2$. The MATLAB® code uses an elliptical parametrization,

$$x = 4\sin\phi\cos\theta, \quad y = 3\sin\phi\sin\theta, \quad x = 2\cos\phi,$$

which bears some similarities to the spherical coordinate transformation (Page 26) of the Cartesian coordinates.

This plot has been made partially transparent using the ``FaceAlpha'' property. The `line` commands are used to set $x$-, $y$-, $z$-axes.

```
theta=linspace(0,2*pi,41); phi=linspace(0,pi,41);
X=4*sin(phi')*cos(theta); Y=3*sin(phi')*sin(theta);
Z=2*cos(phi')*ones(1,41);
surf(X,Y,Z,'FaceAlpha',0.6), hold on
line([-5,0,0;5,0,0],[0,-5,0;0,5,0],[0,0,-5;0,0,5],...
'color','k','linewidth',2)
axis off
```

**Figure 1.28** The graph of $\dfrac{x^2}{4^2} + \dfrac{y^2}{3^2} + \dfrac{z^2}{2^2} = 1$.

# 1.F   Level sets

In many areas of mathematics, physics, and engineering, there arise equations
of the form

$$f(\boldsymbol{x}) = f(x_1, \ldots, x_n) = c$$

where $c$ is a constant real scalar. Although this expression appears to place
interest on the outcome of a function evaluation, it actually raises the question of what values (points) of the argument, $\boldsymbol{x}$, give rise to the specific value
of $f$. This so-called *inverse* problem leads to the notion of a *level set*.

---

**Definition 1.14**
*The set of all points $\boldsymbol{x} \in \mathbb{R}^n$ which give the constant value $c$ for $f$ is called
a **level set**, $L$, or more precisely, the $c$-**level set** of $f$:*

$$L = \{\boldsymbol{x} \in D_f : f(\boldsymbol{x}) = c\}$$

---

Students often confuse a level set with the graph of a function. But as we
have said, this concerns the specific subset of points $\boldsymbol{x} = (x_1, x_2, \ldots) \in D_f$,
rather than what comes out of a function evaluation (except, of course, for
the value $c$!).

■   **Example 1.12:**

In $\mathbb{R}^2$, $\boldsymbol{x} = (x, y)$, and

$$L = \{(x, y) \in D_f : f(x, y) = c\}$$

is a *level curve*. For example, the level set $f(x, y) = x^2 + y^2 = 4$ is the set of
points on the circle in the $xy$-plane (in $\mathbb{R}^2$) with centre $(0, 0)$ and radius 2.
In contrast, the graph of $z = f(x, y)$ is a 3D object in $\mathbb{R}^3$.

In $\mathbb{R}^3$, $\boldsymbol{x} = (x, y, z)$, and

$$L = \{(x, y, z) \in D_f : f(x, y, z) = c\}$$

is a *level surface*. For example, the level set $f(x, y, z) = x^2 + y^2 + z^2 = 4$ is
the set of points on the surface of the sphere in $\mathbb{R}^3$ with centre $(0, 0, 0)$ and
radius 2.

■

By construction, determining the level set from the expression $f(\boldsymbol{x}) = c$ is an inverse problem. Sometimes when $f$ is given explicitly, as in Example 1.12, we are able to "solve" for one variable in terms of the others. In the above 2D example $x^2 + y^2 = 4$, we obtain $y = \sqrt{4 - x^2}$, a semicircle curve passing through $(0, 2)$, and $y = -\sqrt{4 - x^2}$, a semicircle curve passing through $(0, -2)$.

The next example shows how a 3D surface can give rise to level sets in different 2D planes.

■   **Example 1.13:**

Consider the circular paraboloid of Figure 1.29: $S = \{(x, y, z) : z = x^2 + y^2; -1 \le x, y \le 1\}$.

Horizontal level sets occur at fixed values of $z$. The paraboloid is shown in Figure 1.29(a) together with the level sets for $z = r^2$, $r = 0.5, 0.6, 0.7, 0.8, 0.9$.

Vertical level sets occur at fixed values of $x$ or $y$. Shown in Figure 1.29(b) is the level set for $y = 0$.

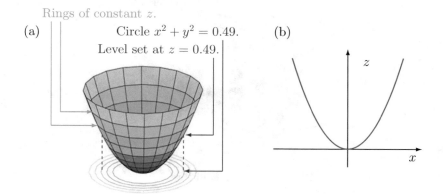

**Figure 1.29**   The paraboloid $z = x^2 + y^2$ (a), with level sets of two kinds in (a) and (b).

The basic figure in (a) was generated using this MATLAB® code:

```
theta=linspace(0,2*pi,21); % 20 intervals around the circle
r=linspace(0,1,11); % 10 intervals from 0 to 1
x=cos(theta')*r; y=sin(theta')*r; z=x.^2+y.^2;
% x, y and z are 21X11 matrices
surf(x,y,z), axis off
hold on % wait for the contours corresponding to
```

```
% the level sets at z=r^2; r=.5, .6, .7, .8, .9
contour(x,y,z,[0.25,0.36,0.49,0.64,0.81])
```

hold off.                                                                        ∎

Example 1.14 draws our attention to the fact that an expression involving only two variables may still describe a function in 3D, although any level sets may have a simple form.

### ∎  Example 1.14:

Consider the parabolic cylinder of Figure 1.30: $S = \{(x,y,z) : z = 4x^2\}$.

Even though there is no specific $y$-dependence, this is a surface in 3D as opposed to the 2D parabola of the last example which we found by setting $y = 0$. The lack of a $y$-dependence means that the shape persists for all values of $y$. The curves of constant $z$ (the level sets) are therefore lines parallel to the $y$-axis.

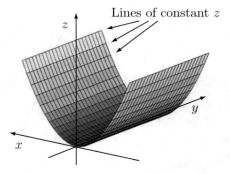

**Figure 1.30**  The graph of $z = 4x^2$.

The figure may be generated using MATLAB® code similar to this:

```
x=linspace(-2,2,25); y=linspace(-2,2,25);
[X,Y]=meshgrid(x,y); Z=4*X.^2;
surf(X,Y,Z)
text(1.0,-2.4,-0.2,'$x$','interpreter','latex','fontsize',32)
text(2.4,1.0,-0.2,'$y$','interpreter','latex','fontsize',32)
text(-2.4,-2.4,14,'$z$','interpreter','latex','fontsize',32)
xticks([]), yticks([]), zticks([0,4,8,12,16])
view(36.5,22)
```

                                                                                 ∎

A level set of the form $g(x, y, z) =$ a constant is equivalent to declaring a function of two variables, which can in principle at least be plotted in a 3D diagram as we see in the next example. Note that the points in this 3D set lie in the domain of $g$, *not the graph of $g$*.

■ **Example 1.15:**

Lastly, consider the surfaces shown in Figure 1.31: $w = g(x, y, z) = z - f(x, y) = z - x^2 + y/5 = k$. Let $k$ take the values $1, 2, 3, 4$.

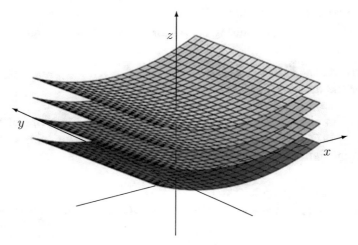

**Figure 1.31**   The level sets of $z - x^2 + y/5 = k$
for $k = 1, 2, 3, 4$.

The MATLAB® code:

```
X=ones(1,21)'*linspace(-1,1,21); Y=X';
Z=zeros(21,21);
for k=1:4
Z=X.^2+k-Y/5;
surf(X,Y,Z), hold on
end
xticks([-1,0,1]); yticks([-1,0,1]); zticks([0,1,2,3,4,5,6]);
text(0.6,-1.2,0,'$x$','interpreter','latex','fontsize',24)
text(-1.2,0.7,-0.1,'$y$','interpreter','latex','fontsize',24)
text(-1.2,1.1,4.5,'$z$','interpreter','latex','fontsize',24)
```

```
axis([-1 1 -1 1 0 6])
hold off
```

∎

🖎   **Mastery Check 1.12:**

Consider the function $f(x, y) = \dfrac{x^3 - y^2}{x^2 - x}$ in the domain

$D_f = \{(x, y) : x \geq 0,\ y \geq 0\}$.

For what values of the constant $k$ can the level set $f(x, y) = k$ be solved for $y$ as a function of $x$ throughout $D_f$? Choose two such values for $k$, and use MATLAB® to plot the resulting curves for $0 \leq x \leq 2$ on the same figure.

🖎

🖎   **Mastery Check 1.13:**

Consider the function $f(x, y, z) = x^2 - y^2 + z^2$ defined for

$D_f = \{(x, y, z) : |x| \leq 2,\ |y| < \infty,\ z \geq 0\}$.

Show that we may solve the level set $f(x, y, z) = k$ for $z$ in terms of $x$ and $y$ for all $k \geq 4$. For the cases $k = 4, 6, 8$, draw the graphs for $|x| \leq 2$, $|y| \leq 2$, on the same figure.

🖎

🖎   **Mastery Check 1.14:**

Consider $F(w, x, y, z) = 36x^2 - 36y^2 - 4z^2 - 9w^2 = 0$. This is a level set in $\mathbb{R}^4$.

Describe in words the precise subsets of this level set that arise from setting $x = 0,\ 1,\ 2,\ 3$.

Use MATLAB® to draw the graphs for cases $x = 1,\ 2$, on the same axes. You may wish to use the ``FaceAlpha`` property (see Page 37).

🖎

Although in this book we shall generally work with problems and examples which admit closed-form inversion of the level set equation, the reader should acknowledge that this will not always be possible in practical situations.

# 1.G  Supplementary problems

**Section** 1.A

1. Suppose three vectors $u$, $v$, $w$ are such that $u + v + w = 0$. Show that $u \times v = v \times w = w \times u$. With the aid of a diagram describe what this result means.

2. Let $x_i = (x_i, y_i, z_i)$, $i = 0, 1, 2, 3$ be four non-coplanar points in $\mathbb{R}^3$, and let vectors $u_i = x_i - x_0$, $i = 1, 2, 3$, be edges of the tetrahedron formed by those points. Consider the four vectors $a, b, c, d$ with magnitudes in turn equal to twice the area of the four faces of the tetrahedron, and directions outwards and normal to those faces. Express these vectors in terms of the $u_i$ and hence show that $a + b + c + d = 0$.
   (In Figure 1.32, regard the point $(x_2, y_2, z_2)$ as being to the rear, without any loss of generality. Normal vectors $a$ and $d$ are shown.)

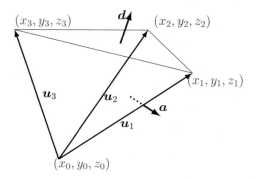

**Figure 1.32**  Four vectors in 3D space.

3. Let $u = (a_1, a_2, a_3)$, $v = (b_1, b_2, b_3)$, and $w = (c_1, c_2, c_3)$ be vectors in 3-D space.

   (a) Show that $u \cdot (v \times w) = (u \times v) \cdot w$. That is, show that in the scalar triple product the "dot" and the "cross" can change places.
   (b) Show that $u \times (v \times w) = (u \cdot w)v - (u \cdot v)w$.

**Section** 1.B

4. Consider the three points $a = (1, -1, 2, 2)$, $b = (3, 1, -1, 1)$, $c = (-2, 0, 2, -1)$ in $\mathbb{R}^4$.

   (a) Find the distances $|b - a|$, $|c - b|$, and $|a - c|$.

   (b) Do either of the points $b$ and $c$ lie inside the open sphere $S_{3\sqrt{2}}(a)$?

   (c) Find the angle $\theta$ between $b - a$ and $c - a$.

5. In $\mathbb{R}^n$, for what values of $k$ is $b = (k, k, k, ...)$ inside $S_r(a)$ when $a = (1, 1, 1, ...)$?

6. Sketch the following regions and determine their boundaries. Also establish whether the regions are open or closed or neither.

   (a) $\{(x, y) : |x| + |y| < 1\}$.

   (b) $\{(x, y) : \max(|x|, |y|) < 1\}$.

   (c) $\{(x, y) : x^2 \leq y \leq \sqrt{x}\}$.

   (d) $\{(x, y) : 1 < (x - 1)^2 + (y + 1)^2 \leq 2\}$.

   (e) $\{(x, y) : |x + 3y| \leq 3\}$.

7. Prove the following statements about sets:

   (a) The boundary of a set $M$ is a closed set.

   (b) A set $M$ is closed $\iff M = \overline{M}$.

   (c) A set $M$ is open $\iff M = \text{Int}(M)$.

   (d) A set $M$ is closed $\iff$ its complement $M^c$ is open.

   (e) The union of any number of open sets is open, and any finite intersection of open sets is open.

   (f) The intersection of any number of closed sets is closed, and any finite union of closed sets is closed.

8. If $x = (x, y, z)$ and $|x| = \sqrt{x^2 + y^2 + z^2}$, show that
$$\max(|x|, |y|, |z|) \leq |x| \leq |x| + |y| + |z| \leq \sqrt{3}|x| \leq 3\max(|x|, |y|, |z|)$$

**Section** 1.C

9. Describe the implied domain $D_f$ for each of the following functions: Is it (i) closed?, (ii) finite?, (iii) compact?

(a) $f(x, y, z) = \arcsin(x^2 + y^2 + z^2)$.

(b) $f(x, y, z) = \arcsin(x^2 + y^2)$.

(c) $f(x, y) = \arctan(x^2 + y^2)$.

(d) $f(x, y, z) = \ln(1 - |x + y + z|)$.

## Section 1.D

10. MATLAB® is able to plot functions expressed in 2D polar coordinates using a plotter called `ezpolar`. Use this function to plot the curves $r = 2\sin n\theta$, $0 \le \theta \le 2\pi$, for $n = 2, 3, 4$, on separate graphs.

11. (a) Express the 2D polar function $r = 2a\sin\theta$, $0 \le \theta \le \pi$, $a$ constant, in Cartesian coordinates, and describe the curve.

   (b) What is the curve defined by $r = 2a\sin\theta$, $0 \le \theta \le 2\pi$?

12. *A curve in $\mathbb{R}^3$ can be sufficiently prescribed in terms of one independent parameter.*

   Represent all points on the curve of intersection of the plane $ax + by + cz = d$ with the unit sphere centred at the point $(x_0, y_0, z_0)$ in terms of spherical coordinates, if the plane passes through the sphere's centre.

   Give conditions that must be satisfied by the constants $a, b, c, d$ for the intersection to be possible and your representation valid.

   Hints: Set the origin for the spherical coordinate system to be at the sphere's centre. Select the longitude angle $\theta$ as the independent variable, then the latitude angle $\phi$ becomes a function of $\theta$.

13. The surfaces $z^2 = 2x^2 + 2y^2$ and $z = y - 1$ intersect to create a closed curve. Use cylindrical coordinates to represent points on this curve.

14. *A surface in $\mathbb{R}^3$ can be sufficiently prescribed in terms of two independent parameters.*

   Represent all points on the plane $ax + by + cz = d$ within and coinciding with the sphere $x^2 + y^2 + z^2 = R^2$ in terms of spherical coordinates defined with respect to the origin at the sphere's centre.

   Give conditions that must be satisfied by the constants $a, b, c, d, R$ for the intersection to be possible and your representation valid.

15 The surfaces $z = 1 + x + y$, $z = 0$, and $x^2 + y^2 = 1$ bound a closed volume, $V$. Represent all points in $V$ and on its boundary in spherical coordinates. Be mindful of the domains of the respective independent variables.

**Section** 1.E

**Figure 1.33** The graph of $\dfrac{x^2}{a^2} + \dfrac{y^2}{b^2} - \dfrac{z^2}{c^2} = 1$.

16. Figure 1.33 shows the elliptic hyperbola of one sheet,

$$\frac{x^2}{a^2} + \frac{y^2}{b^2} - \frac{z^2}{c^2} = 1.$$

Use MATLAB® to reproduce this plot.

17. Use MATLAB® to reproduce Figure 1.34 which shows the graphs of the cylinders $x^2 - 2x + y^2 = 0$ and $z^2 - 2x = 0$ for $0 \le x \le 2$, $-2 \le y \le 2$, $-2 \le z \le 2$, plotted on the same axes.

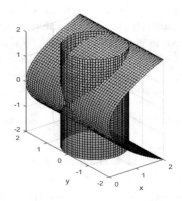

**Figure 1.34** Two intersecting cylinders.

18. Use MATLAB® to draw a sketch graph of the cone $x^2 + y^2 = xz$ for $-4 \le z \le 4$.

19. Describe and sketch the graphs of the following functions using ideas analogous to those described in Example 1.6.

    (a) $f(x,y) = \sqrt{1 - x^2 - y^2}$, for $x^2 + y^2 \le 1$.
    (b) $f(x,y) = \sqrt{1 - y^2}$, for $|y| \le 1, x \in \mathbb{R}$.
    (c) $f(x,y) = 1 - (x^2 + y^2)$, for $(x,y) \in \mathbb{R}^2$.
    (d) $f(x,y) = \dfrac{1 - \sqrt{x^4 + y^4}}{x^2 + y^2 + 1}$, for $(x,y) \in \mathbb{R}^2$.

## Section 1.F

20. Consider the function $f(x,y,z) = \sin(xy) + \cos(yz)$, $-\frac{\pi}{2} \le x, y, z \le \frac{\pi}{2}$.

    Use MATLAB® to draw the graphs of the level sets $x = 1$, $x = 2$; $y = 1$, $y = 2$; and $z = 1$, $z = 2$.

21. Sketch the level curves of the following functions and determine the conditions for the allowed constant values of $f$.

    (a) $f(x,y) = x^2 + y^2 - 4x + 2y$.
    (b) $f(x,y) = x^2 y$.
    (c) $f(x,y) = x\sqrt{y^2 + 1}$.
    (d) $f(x,y) = \dfrac{x}{x^2 + y^2}$.
    (e) $f(x,y) = \dfrac{4x}{1 + x^2 + y^2}$.

# Chapter 2

# Differentiation of multivariable functions

What would calculus be without derivatives? In this chapter we cover the theory of the differential calculus, beginning with the limit concept as it pertains to functions of many variables and their derivatives. Considerable emphasis is placed on the geometric meaning of partial derivatives and of differentiability in general. The discussion also covers higher-order derivatives and introduces the new concept of the *gradient* of a function.

The focus of attention is then directed to composite functions, the chain rule of partial differentiation, and to implicit functions. The dedication of considerable space to these latter topics is motivated partly by their level of complication which is much greater than in the case of functions of one variable, and partly by the simple fact that they are commonly encountered in practice.

Having established these foundation concepts we shall put them to practical use in Chapter 3, where we discuss a fair assortment of applications.

## 2.A  The derivative

Differentiation is all about *limiting processes* and *linear approximations*. To prepare us for that discussion we turn to the 1D case for inspiration. Many of the necessary basic features and results of limits of functions of one variable appearing below were covered in Chapter 1. The reader may wish to refer back to Section 1.C for details.

© Springer Nature Switzerland AG 2020                                          49
S. J. Miklavcic, *An Illustrative Guide to Multivariable and Vector Calculus*,
https://doi.org/10.1007/978-3-030-33459-8_2

**Definition 2.1**

*Let $f : \mathbb{R} \longrightarrow \mathbb{R}$ be a continuous function over an open interval domain. That is, in more mathematical notation, let $f \in C(D_f)$, where the domain of $f$ $D_f \subset \mathbb{R}$ is open.*

*Let $x_0$ and $x_0 + h \in D_f$. If $\lim\limits_{h \to 0} \dfrac{f(x_0 + h) - f(x_0)}{h}$ exists, it is called the* **(first) derivative of** $f$ **at** $x_0$, *and we write*

$$\left.\frac{dy}{dx}\right|_{x_0} = f'(x_0) = \lim_{h \to 0} \frac{f(x_0 + h) - f(x_0)}{h},$$

*where $\left.\dfrac{dy}{dx}\right|_{x_0}$ is the* **slope** *of the tangent line to the* **graph** *of $f$ at the point $\big(x_0, y_0 = f(x_0)\big)$.*

*An equivalent definition:*

$$\left.\frac{dy}{dx}\right|_{x_0} = \lim_{x \to x_0} \frac{f(x_0) - f(x)}{x_0 - x}; \quad x_0, x \in D_f.$$

Note that either explicitly or implicitly we have assumed the following properties which are essential criteria for the existence of the limit:

(i) $x_0, x_0 + h \in D_f$

— both points belong to the domain, $D_f$

(ii) $\lim\limits_{h \to 0^-} f(x_0 + h) = \lim\limits_{h \to 0^+} f(x_0 + h) = f(x_0)$

— the left limit equals the right limit which equals the function value at $x_0$

Thus, $\lim\limits_{h \to 0} f(x_0 + h)$ exists and is equal to $f(x_0)$.

(iii) Similarly, $\lim\limits_{h \to 0^-} \dfrac{f(x_0 + h) - f(x_0)}{h} = \lim\limits_{h \to 0^+} \dfrac{f(x_0 + h) - f(x_0)}{h}$

— the left and right limits of these ratios exist and are equal

The reasons why these conditions are essential for the definition of a derivative are demonstrated in the following two classic examples of problem cases.

## ■ Example 2.1:

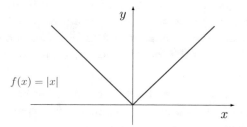

$$f(x) = |x|$$

**Figure 2.1** A function not everywhere differentiable.

For the function shown in Figure 2.1, (i) and (ii) are satisfied everywhere. At $x = 0$, however, although the left and right limits of (iii) exist, they are not equal, implying that no derivative exists there. Everywhere else (iii) is satisfied.

■

## ■ Example 2.2:

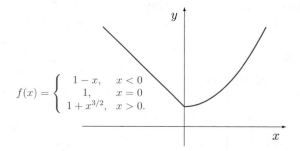

$$f(x) = \begin{cases} 1 - x, & x < 0 \\ 1, & x = 0 \\ 1 + x^{3/2}, & x > 0. \end{cases}$$

**Figure 2.2** Another function not everywhere differentiable.

For the function shown in Figure 2.2, the only problem appears at $x = 0$. Conditions (i) and (ii) are satisfied, but in the case of condition (iii) we have that

$$\lim_{h \to 0^-} \frac{f(0+h) - f(0)}{h} = \lim_{h \to 0^-} \frac{(1-h) - 1}{h} = -1.$$

$$\lim_{h \to 0^+} \frac{f(0+h) - f(0)}{h} = \lim_{h \to 0^+} \frac{(1 + h^{3/2}) - 1}{h} = \lim_{h \to 0^+} h^{1/2} = 0.$$

That is, the left limit is not equal to the right limit, implying that no derivative exists there.

■

Point (ii) is the definition of continuity at $x_0$ (Section 1.C). A function for which any of the equalities is not satisfied is said to be discontinuous at $x_0$. What we are now saying is that continuity is a necessary but not a sufficient condition for differentiability. Functions for which (iii) is not satisfied at any point, $x_0$, such as those of the foregoing examples, are said to be *singular* at that point.

Now let us apply what we have learnt for a function of one variable to the case of a function of two variables. The most obvious analogous expression of a limit generalized to some function $f : \mathbb{R}^2 \longrightarrow \mathbb{R}$ of two variables is:

$$\lim_{P_1 \to P_0} \frac{f(x_0, y_0) - f(x_1, y_1)}{\sqrt{(x_1 - x_0)^2 + (y_1 - y_0)^2}}. \tag{2.1}$$

If this limit exists, should we call it "the" derivative of $f$? Alongside this question we also need to ask what are the generalizations of criteria (i)–(iii) to $\mathbb{R}^2$ (or $\mathbb{R}^3$, … , $\mathbb{R}^n$)?

The graphical foundation for the limit expression (2.1) is shown in Figure 2.3. The things to note are, firstly, the graph of $f$ is suspended in 3D; secondly, the domain $D_f$ lies in the $xy$-plane; thirdly, the points $P_0$ and $P_1$ in $D_f$ give rise to values $z_0$ and $z_1$, respectively; and finally, the line in the domain joining $P_0$ and $P_1$ traces out the black curve in the graph of $f$.

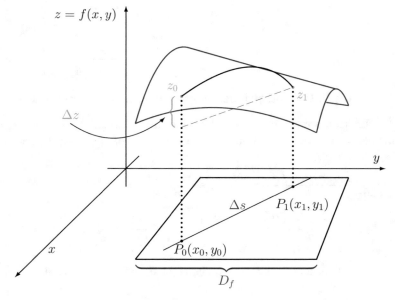

**Figure 2.3** A 3D graph of a function of two variables.

In direct analogy with the 1D derivative we have $\Delta z = f(x_0, y_0) - f(x_1, y_1)$, and $\Delta s = \sqrt{\Delta x^2 + \Delta y^2}$.

Referring to Figure 2.3, for the limit process to make sense several related questions need to be addressed:

* does $f(x_0, y_0)$ exist, or equivalently, does $(x_0, y_0)$ belong to $D_f$?

* is $f$ continuous at $(x_0, y_0)$?

* does the limit $\lim\limits_{P_1 \to P_0}$ exist if $P_1 \in D_f$?

Over the next few pages we consider these questions with the aim of establishing a set of conditions for the existence of derivatives and possibly a set of guidelines that can be followed in applications.

## 2.B   Limits and continuity

In Chapter 1 the pointwise limit of a function of one variable was explained in some detail, culminating in the $\epsilon$–$\delta$ condition shown in Figure 1.16. The latter condition permits a direct generalization of the limit concept to functions of several variables. Keep in mind that the discussion below pertains to given points in $\mathbb{R}^n$, *i.e.* it too holds pointwise.

---

**Definition 2.2**

*A function $f : \mathbb{R}^n \longrightarrow \mathbb{R}$ approaches a limit $L$ as $\boldsymbol{x} \to \boldsymbol{a}$ for all points $\boldsymbol{x}$ in a **neighbourhood** of $\boldsymbol{a}$ belonging to $D_f$ if:*

*Given any small positive number $\epsilon$, another positive number $\delta$, which may depend on $\epsilon$, can be found such that if $\boldsymbol{x}$ is within a radius $\delta$ of $\boldsymbol{a}$, then $f$ will be within a radius $\epsilon$ of $L$.*

*In mathematical notation: Given any $\epsilon > 0$ there exists a $\delta > 0$ such that wherever $0 < |\boldsymbol{x} - \boldsymbol{a}| < \delta$ then $|f(\boldsymbol{x}) - L| < \epsilon$.*

*We write $\lim\limits_{\boldsymbol{x} \to \boldsymbol{a}} f(\boldsymbol{x}) = L$.*

---

If it can be established that two functions, $f$ and $g$, satisfy the conditions of this fundamental definition at a given point, then a number of important results involving their combination follow. We refer to these as *laws* and apply them to all well-behaved functions.

**Limit laws:** If $\lim\limits_{x \to a} f(x) = L, \quad \lim\limits_{x \to a} g(x) = M,$ then the following sum, product, quotient, convergence and composition results can be proved.

(a) $\lim\limits_{x \to a} \big(f(x) + g(x)\big) = L + M$

(b) $\lim\limits_{x \to a} \big(f(x).g(x)\big) = L.M$

(c) $\lim\limits_{x \to a} \dfrac{f(x)}{g(x)} = \dfrac{L}{M} \quad (M \neq 0)$

(d) $\lim\limits_{x \to a} f(x) = \lim\limits_{x \to a} g(x)$ and $f(x) \leq h(x) \leq g(x)$ means that $\lim\limits_{x \to a} h(x)$ exists and equals $L$ which equals $M$ (a "squeeze theorem").

(e) If $F(t)$ is a continuous function at $t = L$ then

$$\lim\limits_{x \to a} F\big(f(x)\big) = F(L) = F\big(\lim\limits_{x \to a} f(x)\big).$$

That is, for continuous functions, we may interchange the limit and function composition operations.

## ■ Example 2.3:

Here is a proof of limit law (a) using the $\epsilon$–$\delta$ concept in Definition 2.2.

We may assume that, given $\epsilon_1 > 0$ and $\epsilon_2 > 0$, we have found numbers $\delta_1 > 0$ and $\delta_2 > 0$ such that $|f(x) - L| < \epsilon_1$ whenever $|x - a| < \delta_1$, and $|g(x) - M| < \epsilon_2$ whenever $|x - a| < \delta_2$.

For given arbitrarily small $\epsilon > 0$, let $\epsilon_1 = \epsilon_2 = \frac{1}{2}\epsilon$. Then we have

$$\big|\big(f(x) + g(x)\big) - \big(L + M\big)\big| = \big|\big(f(x) - L\big) + \big(g(x) - M\big)\big|$$
$$\leq |f(x) - L| + |g(x) - M|$$
$$\text{by the triangle inequality,}$$
$$< \epsilon_1 + \epsilon_2 = \epsilon \quad \text{provided both}$$
$$|x - a| < \delta_1 \text{ and } |x - a| < \delta_2.$$

Now we may choose $\delta = \min(\delta_1, \delta_2)$, and we then have

$$\big|\big(f(x) + g(x)\big) - \big(L + M\big)\big| < \epsilon \text{ whenever } |x - a| < \delta.$$

Thus, we have proved that $\lim\limits_{x \to a} \big(f(x) + g(x)\big) = L + M.$ ■

## ✍  Mastery Check 2.1:
Prove the limit laws (b)–(e).

Hint: For law (b), with the assumptions in Example 2.3, assume
$|\boldsymbol{x} - \boldsymbol{a}| < \delta = \min(\delta_1, \delta_2)$, and write $f(\boldsymbol{x}) = L + e_1(\boldsymbol{x})$, which implies
$|e_1(\boldsymbol{x})| < \epsilon_1$, and similarly for $g$. Expand $f(\boldsymbol{x}).g(\boldsymbol{x})$ in terms of $e_1$ and $e_2$.
Let $\epsilon_1 = \frac{1}{3}\epsilon/|M|$ and $\epsilon_2 = \frac{1}{3}\epsilon/|L|$ and consider $|f(\boldsymbol{x}).g(\boldsymbol{x}) - L.M| < \epsilon$.
For law (c), prove first that $\lim\limits_{x \to a} \dfrac{1}{g(\boldsymbol{x})} = \dfrac{1}{M}$ and then invoke law (b).

✍

## ■  Example 2.4:
Consider $L = \lim\limits_{(x,y) \to (1,\pi)} \dfrac{\cos(xy)}{1 - x - \cos y} = \lim\limits_{(x,y)} \dfrac{g(x,y)}{h(x,y)}$, noting in particular that
$\lim\limits_{(x,y) \to (1,\pi)} h(x,y) \neq 0$. Applying the standard rules we find that

$$L = \frac{\lim \cos(xy)}{\lim(1 - x - \cos y)} = \frac{-1}{+1} = -1.$$

Here, we have used the sum, product, quotient, and composition laws.

■

In evaluating limits of any well-behaved $f : \mathbb{R}^n \longrightarrow \mathbb{R}$ for $n > 2$, we follow
the exact same process as implied in the above example: besides using the
limit laws, the reader can also make use of results from the study of limits
of functions of one variable, some of which are listed on Page 24. However,
the simple statement made in the limit definition hides considerable detail
that we need to confront in more complicated cases. Definition 2.2 implicitly
means that

* $\lim\limits_{x \to a} f(\boldsymbol{x})$ exists and is equal to $L$ if $f \longrightarrow L$ *independently* of how $\boldsymbol{x}$
  approaches $\boldsymbol{a}$!

* The limit $L$, if it exists, is unique!

* *No* limit of $f$ exists if $f$ has different limits when $\boldsymbol{x}$ approaches $\boldsymbol{a}$ along
  different curves!

The graphical depiction of Definition 2.2, in analogy with Figure 1.16, is
shown in Figure 2.4 on the next page.

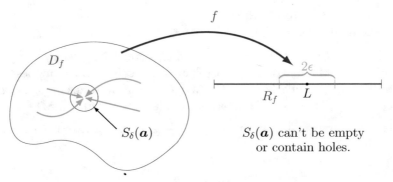

**Figure 2.4** Schematic of the limit definition in 2D.

As in the 1D cases (Examples 2.1 and 2.2), the idea of considering limit values from multiple approaches is best illustrated by an example which fails to meet one or more criteria, such as the example below.

■ **Example 2.5:**

Consider $\lim\limits_{(x,y)\to(1,1)} \dfrac{x-y}{x-1}$. (This is of the form $\dfrac{0}{0}$. The function is therefore undefined at $(1,1)$.) We attempt to evaluate the limit by approaching the point $(1,1)$ along four different paths as shown in Figure 2.5.

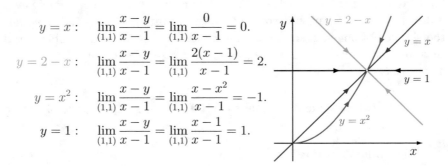

**Figure 2.5** Different paths of approach to a limit point.

The resulting limiting values found by following these different paths are all different. From this we conclude that no limit exists. Note that it is enough for *one* of these cases to give a different result for us to conclude that no limit exists. The graph of the function is shown in Figure 2.6.

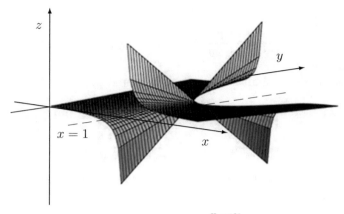

**Figure 2.6**  The graph of $z = \dfrac{x-y}{x-1}$, $0 \le x, y \le 2$.

The point $(1,1)$ is the cusp where the two sheets of the function meet (along the dashed line).

■

✍  **Mastery Check 2.2:**

Consider the function $f(x,y) = \dfrac{x^2 + 2y^2}{2x^2 + y^2}$. Does $\displaystyle\lim_{(x,y)\to(0,0)} f(x,y)$ exist? Hint: Note that $f(x,y)$ is of the form $\dfrac{0}{0}$ at $(0,0)$ making $f(x,y)$ undefined at $(0,0)$. Take limits along the lines $x = 0$, $y = x$, and finally along the line $y = kx$, and see if your limits agree.

✍

From the perspective of effort expended, cases where a limit fails to exist are most often the least taxing, a few well-chosen approach paths will suffice. Now we need to ask, what about the cases where we get the same result for a few different trials? Do we need to try all the infinite number of approach directions to be convinced? One solution is proposed in the next example.

### ■ Example 2.6:

Consider $\displaystyle\lim_{(x,y)\to(0,0)} \frac{x^3 - x^2 y}{x^2 + y^2 + xy}$. Note once again that the function is undefined at the origin.

First we evaluate the limit along a few simple paths.

$$\text{Along } y = 0: \ \lim_{(0,0)} \frac{x^3 - x^2 y}{x^2 + y^2 + xy} = \lim_{(0,0)} \frac{x^3}{x^2} = \lim_{x\to 0} x = 0.$$

$$\text{Along } x = 0: \ \lim_{(0,0)} \frac{x^3 - x^2 y}{x^2 + y^2 + xy} = \lim_{y\to 0} \frac{0}{y^2} = 0.$$

We get the same result along any straight line $y = kx$. If the limit exists, it must be 0!

So, consider an arbitrary curve $r = f(\theta) > 0$, where $x = r\cos\theta$, $y = r\sin\theta$, and let $r \to 0$. (Shown in Figure 2.7 is one of the cases $r \to 0$ as $\theta$ increases, but any path in the plane will do.)

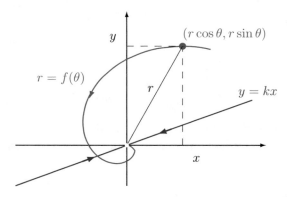

**Figure 2.7** Example of conventional and unconventional paths to a limit point.

Substitute the polar functions for $x$ and $y$ in the definition of the function limit.

$$\left| \frac{x^3 - x^2 y}{x^2 + y^2 + xy} - 0 \right| = \left| \frac{r^3 \cos^3\theta - r^3 \cos^2\theta \sin\theta}{r^2 \cos^2\theta + r^2 \cos^2\theta + r^2 \cos\theta \sin\theta} \right|$$

$$= \left| \frac{r^3 \cos^2\theta(\cos\theta - \sin\theta)}{r^2(1 + \cos\theta\sin\theta)} \right|$$

$$= r\cos^2\theta \, \frac{|\cos\theta - \sin\theta|}{|1 + \sin\theta\cos\theta|}$$

(It is actually sufficient to stop here: the denominator is not zero and the numerator is bounded and proportional to $r$ which $\to 0$.)

$$\left| \frac{x^3 - x^2 y}{x^2 + y^2 + xy} - 0 \right| = r \cos^2 \theta \frac{\sqrt{2}|\cos \theta \cos(\pi/4) - \sin \theta \sin(\pi/4)|}{|1 + \frac{1}{2} 2 \sin \theta \cos \theta|}$$

$$\leq r\sqrt{2} \frac{|\cos(\theta + \pi/4)|}{|1 + \frac{1}{2}\sin(2\theta)|}$$

$$\leq r \frac{\sqrt{2}}{1/2} = 2\sqrt{2} r \longrightarrow 0 \text{ as } r \to 0.$$

In these steps we have used only known properties of the trigonometric functions for arbitrary angles.

Thus, given $\epsilon > 0$, however, small, we can find a $\delta$, a function of $\epsilon$ $\left( \text{choose } \delta = \dfrac{\epsilon}{2\sqrt{2}} \right)$, such that

$$\left| \frac{x^3 - x^2 y}{x^2 + y^2 + xy} - 0 \right| < \epsilon \quad \text{whenever} \quad r < \delta.$$

Given that we have invoked an arbitrary curve whose sole requirement is to pass through the limit point (the origin) the result is general, the limit exists, and is indeed 0. The surface itself is reproduced in Figure 2.8.

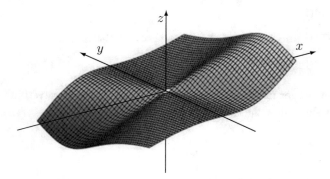

**Figure 2.8**  The graph of $z = \dfrac{x^3 - x^2 y}{x^2 + y^2 + xy}$.

✍ **Mastery Check 2.3:**

Consider the function $f(x, y) = \dfrac{x + y}{\ln(x^2 + y^2)}$ (which is undefined at $(0, 0)$).

Does $\lim\limits_{(x,y)\to(0,0)} f(x, y)$ exist?

Hint: Use the approach featured in Example 2.6, and standard limits.

✍

Not all limits behave nicely!

Up to this point, the student may be inclined to think that taking limits along lines $y = kx$, $0 \le k \le \infty$, is sufficient to determine whether a limit at $(x, y) = (0, 0)$ exists or not. That is, that limit operations in $\mathbb{R}^2$ are a straightforward (no pun intended) extensions of the essential criteria listed on Page 50 for limits in $\mathbb{R}$.

But this is not so! The following case is a counterexample to Example 2.6:

✍ **Mastery Check 2.4:**

Consider the function $f : \mathbb{R}^2 \longrightarrow \mathbb{R}$ defined by

$$f(x, y) = \frac{x^4 y^2}{(x^4 + y^2)^2}, \quad (x, y) \ne (0, 0).$$

Your task is to show that $\lim\limits_{(x,y)\to(0,0)} f(x, y)$ does not exist (without first drawing the graph).

Hint: Consider another class of curves through the origin.

✍

The student reader might get a better idea of all that is involved in evaluating limits with the following summary flowchart.

**Flowchart 2.1:   How to work through a limit problem**

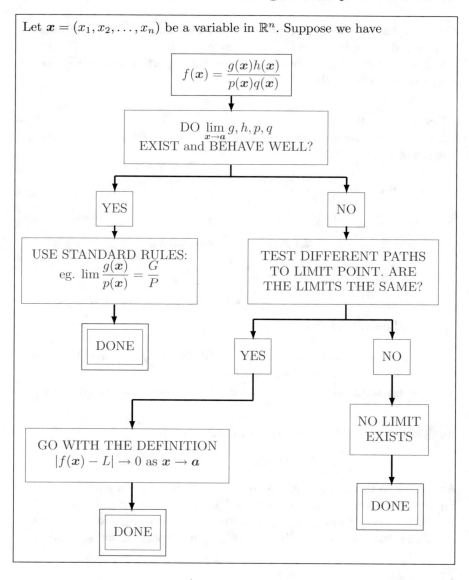

## 2.C Partial derivatives

Once the intricacies of limit processes for functions of many variables are understood, the application of these same principles to *ratios of differences* is relatively straightforward. In fact, the concept of partial derivatives becomes a simple extension of the 1D ordinary derivative.

---

**Definition 2.3**
*Let $\boldsymbol{x}_0$ be an **interior** point or a **boundary** point of $D_f$ of a continuous function $f : \mathbb{R}^n \longrightarrow \mathbb{R}$.*

* *If the limit* $\displaystyle\lim_{h\to 0} \frac{f(\boldsymbol{x}_0 + h\boldsymbol{e}_j) - f(\boldsymbol{x}_0)}{h}$

$$= \lim_{h\to 0} \frac{f(x_{0,1}, \ldots, x_{0,j} + h, \ldots, x_{0,n}) - f(x_{0,1}, \ldots, x_{0,n})}{h}$$

*exists, we call it the first **partial derivative** of $f$ w.r.t. $x_j$ at the point $\boldsymbol{x}_0$ and denote it $\dfrac{\partial f}{\partial x_j}(\boldsymbol{x}_0)$.*

* *If all $\dfrac{\partial f}{\partial x_j}(\boldsymbol{x}_0)$, $j = 1, \ldots, n$, exist then we say that $f$ is **partially differentiable** at $\boldsymbol{x}_0$.*

* *We say $f$ is **partially differentiable** in $D_f$ if it is partially differentiable at **every** point $\boldsymbol{x}_0 \in D_f$.*

---

Equivalent notations for partial derivatives are:

$$\frac{\partial f}{\partial x_j}(\boldsymbol{x}_0), \quad D_j f(\boldsymbol{x}_0), \quad f_{x_j}(\boldsymbol{x}_0), \quad f_j(\boldsymbol{x}_0).$$

Analogous to the 1D case, functions for which any of the $n$ partial derivative limits fails to exist at a point are said to be singular at that point. That is, a multivariable function may be continuous everywhere in its domain of definition, but need not be differentiable at every point in its domain.

### ✎ Mastery Check 2.5:

Let $f(x, y, z) = xy + z\sin(yz)$. Using Definition 2.3, determine $\dfrac{\partial f}{\partial y}$ at an arbitrary point $(x, y, z)$.

Hint: You may need to use standard limits (see Page 24). ✎

In solving this Mastery Check problem you will have noticed that you could have and would have arrived at the same result had you used the rules of differentiation for functions of one variable, provided you treated $x$ and $z$ as if they were constants! In actual fact, Definition 2.3 effectively states that in taking the limit with respect to one variable, we do keep *all other* variables fixed. It should not come as a surprise that we find this equivalence. We demonstrate this very convenient operational equivalence with an example and leave it to Mastery Check 2.6 to reinforce the procedure.

■  **Example 2.7:**

Let $f(x, y, z) = \ln(1 + e^{xyz}) = g\big(h(x, y, z)\big)$. We wish to calculate $\dfrac{\partial f}{\partial x}, \dfrac{\partial f}{\partial y}, \dfrac{\partial f}{\partial z}$.

In each case we assume two variables are constant and differentiate w.r.t. the third using the chain rule of single-variable calculus:

$$\frac{\partial f}{\partial x} = \frac{dg}{dh}\frac{\partial h}{\partial x} = \frac{1}{1 + e^{xyz}}\, yze^{xyz},$$

$$\frac{\partial f}{\partial y} = \frac{1}{1 + e^{xyz}}\, xze^{xyz},$$

$$\frac{\partial f}{\partial z} = \frac{1}{1 + e^{xyz}}\, xye^{xyz}.$$

■

✎  **Mastery Check 2.6:**

Find the (first-order) partial derivatives of the following functions with respect to each variable:

1. $f(x, y, z) = \dfrac{x^2 + y^2}{x^2 - z^2}$;

2. $f(x, y, u, v) = x^2 \sin(2y) \ln(2u + 3v)$;

3. $f(s, t, u) = \sqrt{s^2 t + stu + tu^2}$;

4. $f(x, y, z) = y \sin^{-1}(x^2 - z^2)$;

5. $f(x, y, z, u) = \sin(3x)\cosh(2y) - \cos(3z)\sinh(2u)$;

6. $f(u, v, w) = u^2 e^{uv^2 w}$.

✎

Now that we can evaluate them, what are partial derivatives?

Let's look more closely at Figure 2.3 (Page 52). Given the foregoing discussion and particularly Definition 2.3 we consider two specific cases of that graph of the function of two variables.

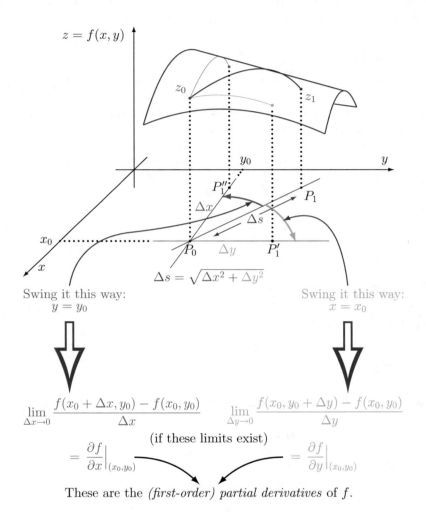

$$\lim_{\Delta x \to 0} \frac{f(x_0 + \Delta x, y_0) - f(x_0, y_0)}{\Delta x} \qquad \lim_{\Delta y \to 0} \frac{f(x_0, y_0 + \Delta y) - f(x_0, y_0)}{\Delta y}$$

(if these limits exist)

$$= \frac{\partial f}{\partial x}\Big|_{(x_0, y_0)} \qquad\qquad = \frac{\partial f}{\partial y}\Big|_{(x_0, y_0)}$$

These are the *(first-order) partial derivatives* of $f$.

**Figure 2.9** Partial derivatives of a function of two variables.

From Figure 2.9 we deduce geometric interpretations for $\dfrac{\partial f}{\partial x}\Big|_0$, $\dfrac{\partial f}{\partial y}\Big|_0$:

$\dfrac{\partial f}{\partial x}\Big|_0$    — the *slope* of the *tangent line* $L_1$ to the curve $f(x, y_0)$ at $(x_0, y_0)$.

$\dfrac{\partial f}{\partial y}\Big|_0$    — the *slope* of the *tangent line* $L_2$ to the curve $f(x_0, y)$ at $(x_0, y_0)$.

In fact, by taking two orthogonal cross sections through the point $(x_0, y_0, f(x_0, y_0))$ on the graph of $f$, one parallel to the $x$-axis and one parallel to the $y$-axis, the following facts can be obtained.

In the cross section parallel to the $xz$-plane, a vector parallel to line $L_1$ is $\boldsymbol{v}_1$ shown in Figure 2.10.

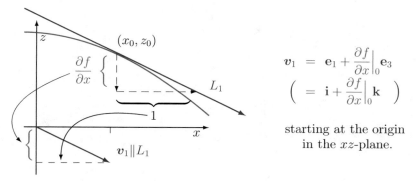

$$\boldsymbol{v}_1 = \mathbf{e}_1 + \frac{\partial f}{\partial x}\Big|_0 \mathbf{e}_3$$
$$\left( = \mathbf{i} + \frac{\partial f}{\partial x}\Big|_0 \mathbf{k} \right)$$

starting at the origin
in the $xz$-plane.

**Figure 2.10**  A tangent vector and line in the $x$-direction.

In the cross section parallel to the $yz$-plane, a vector parallel to line $L_2$ is $\boldsymbol{v}_2$ shown in Figure 2.11.

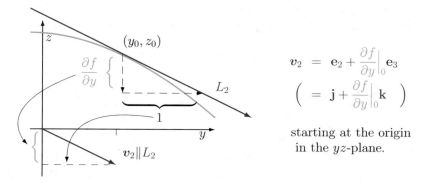

$$\boldsymbol{v}_2 = \mathbf{e}_2 + \frac{\partial f}{\partial y}\Big|_0 \mathbf{e}_3$$
$$\left( = \mathbf{j} + \frac{\partial f}{\partial y}\Big|_0 \mathbf{k} \right)$$

starting at the origin
in the $yz$-plane.

**Figure 2.11**  A tangent vector and line in the $y$-direction.

In the full 3D graph the tangent lines $L_1$ and $L_2$, and corresponding vectors $\boldsymbol{v}_1$ and $\boldsymbol{v}_2$, appear as in the figure below. (Compare the corresponding lines in Figure 2.12 below with those in Figures 2.9, 2.10, and 2.11 above.)

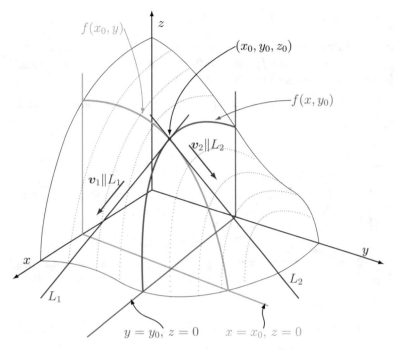

**Figure 2.12**  Tangent vectors and lines to a 3D function graph.

Notice that $\boldsymbol{v}_1$ and $\boldsymbol{v}_2$ are not parallel to each other! This is useful as the vectors $\boldsymbol{v}_1$ and $\boldsymbol{v}_2$ define a *tangent plane*, $T$, with normal vector $\boldsymbol{n}$:

$$
\boldsymbol{n} = \boldsymbol{v}_1 \times \boldsymbol{v}_2 =
\begin{vmatrix}
\mathbf{e}_1 & \mathbf{e}_2 & \mathbf{e}_3 \\
1 & 0 & \dfrac{\partial f}{\partial x}\Big|_0 \\
0 & 1 & \dfrac{\partial f}{\partial y}\Big|_0
\end{vmatrix}
= -\frac{\partial f}{\partial x}\Big|_0 \mathbf{e}_1 - \frac{\partial f}{\partial y}\Big|_0 \mathbf{e}_2 + \mathbf{e}_3.
$$

The plane defined by $L_1$ and $L_2$ is *tangent* to the surface $z = f$ at $(x_0, y_0, z_0)$, is spanned by $\boldsymbol{v}_1$ and $\boldsymbol{v}_2$, and, of course, has the same normal as the normal to the graph of $z = f(x, y, z)$ at $(x_0, y_0, z_0)$.

The equation of this tangent plane can be found from the scalar vector product (Page 4):

$$\boldsymbol{n} \cdot (\boldsymbol{x} - \boldsymbol{x}_0) = 0.$$

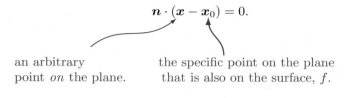

an arbitrary              the specific point on the plane
point *on* the plane.      that is also on the surface, $f$.

$$\implies \quad \frac{\partial f}{\partial x}\Big|_0 (x - x_0) + \frac{\partial f}{\partial y}\Big|_0 (y - y_0) - (z - z_0) = 0$$

$$\implies \quad z - z_0 = \frac{\partial f}{\partial x}\Big|_0 (x - x_0) + \frac{\partial f}{\partial y}\Big|_0 (y - y_0).$$

✍   **Mastery Check 2.7:**

Let $z = f(x, y) = \arcsin(xy)$.

Find the normal vector to the surface generated by $f(x, y)$, and the equation of the tangent plane, at $(1, \frac{1}{2}, \frac{\pi}{6})$.

✍

# 2.D   Differentiability of $f : \mathbb{R}^n \longrightarrow \mathbb{R}$

We can now use the developments of the last section to establish a convenient definition of differentiability, extending the following geometric argument from single-variable calculus.

Consider the function $f : \mathbb{R} \longrightarrow \mathbb{R}$. In saying that $f$ is differentiable at a point $x_0 \in D_f$ we mean, *geometrically*, on the one hand, that there exists a tangent line to $f(x)$ at the point $x_0$ (Figure 2.13):

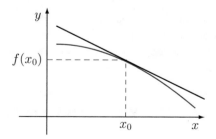

**Figure 2.13**   Tangent line to the graph of $f(x)$.

and *analytically* on the other hand:

$$\lim_{x \to x_0} \frac{f(x) - f(x_0)}{x - x_0} = c \quad \left( = \frac{\mathrm{d}f}{\mathrm{d}x}\Big|_0 \right)$$

$$\Longleftrightarrow \left| \frac{f(x) - f(x_0) - c(x - x_0)}{x - x_0} \right| \longrightarrow 0 \quad \text{as } x \to x_0$$

$$\Longleftrightarrow \frac{f(x) - f(x_0) - c(x - x_0)}{x - x_0} = \rho(x - x_0) \longrightarrow 0 \quad \text{as } x \to x_0.$$

The variable $\rho$ is a function of $(x - x_0)$ which $\longrightarrow 0$ as $x \to x_0$.

Rewriting this last result, we conclude that $f(x)$ is differentiable at $x_0$

$$\Longleftrightarrow \quad f(x) = f(x_0) + c(x - x_0) + |\Delta x|\rho(\Delta x) \text{ and } \lim_{\Delta x \to 0} \rho(\Delta x) = 0$$
$$\Longleftrightarrow \quad f \text{ can be approximated by a line.}$$

The generalization of this argument to a function of two variables $f : \mathbb{R}^2 \longrightarrow \mathbb{R}$ is reasonably straightforward with the tangent line approximation being replaced by a tangent plane approximation.

We say $f(x, y)$ is differentiable at $(x_0, y_0)$

$$\Longleftrightarrow \quad f(x, y) = f(x_0, y_0) + c_1 \Delta x + c_2 \Delta y + |\boldsymbol{\Delta x}|\rho(\boldsymbol{\Delta x})$$
$$\Longleftrightarrow \quad f \text{ can be approximated by a plane at } (x_0, y_0).$$

We formalize this reasoning in an even more general definition for a function of $n$ variables.

---

**Definition 2.4**

*Let $f : \mathbb{R}^n \longrightarrow \mathbb{R}$. $f$ is said to be **differentiable** at $\boldsymbol{x}_0 \in D_f$ if there exists a **linear approximation**, $\phi(\boldsymbol{x} - \boldsymbol{x}_0)$, such that*

$$f(\boldsymbol{x}) = f(\boldsymbol{x}_0) + \phi(\boldsymbol{x} - \boldsymbol{x}_0) + |\boldsymbol{\Delta x}|\rho(\boldsymbol{\Delta x}; \boldsymbol{x}, \boldsymbol{x}_0) \quad \text{near } \boldsymbol{x}_0$$

*with $\phi(\boldsymbol{x} - \boldsymbol{x}_0) = c_1(x_1 - x_{0,1}) + \cdots + c_n(x_n - x_{0,n})$ and $\lim\limits_{\boldsymbol{\Delta x} \to 0} \rho(\boldsymbol{\Delta x}; \boldsymbol{x}, \boldsymbol{x}_0) = 0$. A function for which no linear approximation can be defined at a point is said to be **singular** at that point.*

---

Although it is not critical to the discussion here (instead see Section 2.I and Chapter 3), a word or two about the function $\rho$ (relevant to both the 1D and nD cases) is warranted.

The condition that the function $f$ is differentiable at $\boldsymbol{x}_0$ is equivalent to the condition of the existence of a tangent plane, $\phi$, at that point. For $\boldsymbol{x} \neq \boldsymbol{x}_0$, a rearrangement of the equation in Definition 2.4 then defines the function $\rho(\boldsymbol{\Delta x}; \boldsymbol{x}, \boldsymbol{x}_0)$ as the ratio of the difference (effectively) of $f$ and $\phi$ to $|\boldsymbol{\Delta x}|$,

$$\rho(\boldsymbol{\Delta x}; \boldsymbol{x}, \boldsymbol{x}_0) = \frac{f(\boldsymbol{x}) - f(\boldsymbol{x}_0) - \phi(\boldsymbol{x} - \boldsymbol{x}_0)}{|\boldsymbol{\Delta x}|},$$

which should only be a nonlinear contribution. Definition 2.4 then states that as a further condition for differentiability, this function must vanish in the limit $|\boldsymbol{\Delta x}| \to 0$. Essentially, for differentiability $f(\boldsymbol{x}) - f(\boldsymbol{x}_0)$ must behave as a linear function of the independent variables. We clarify this explanation with an example.

■ **Example 2.8:**
Consider the function $f(x, y) = xy^2$ and the point $\boldsymbol{x}_0 = (-2, 1)$. We introduce $\boldsymbol{h} = (h, k)$ so that $\boldsymbol{x} = \boldsymbol{x_0} + \boldsymbol{h} = (-2 + h, 1 + k)$. Then

$$\begin{aligned} f(\boldsymbol{x}_0 + \boldsymbol{h}) - f(\boldsymbol{x}_0) &= (-2 + h)(1 + k)^2 - (-2)1^2 \\ &= (-2 + h)(1 + 2k + k^2) + 2 \\ &= (h - 4k) + (-2k^2 + 2hk + hk^2). \end{aligned}$$

In the last expression on the right-hand side, the two pairs of parentheses separate the linear approximation, $\phi(h, k) = h - 4k$, from the remaining nonlinear terms. From the latter terms we then form our $\rho$ function

$$\rho(\boldsymbol{h}; \boldsymbol{x}_0 + \boldsymbol{h}, \boldsymbol{x}_0) = \frac{-2k^2 + 2hk + hk^2}{\sqrt{h^2 + k^2}}$$

which vanishes in the limit $|\boldsymbol{h}| = \sqrt{h^2 + k^2} \to 0$ since the numerator is at least quadratic in $h$ and $k$ while the denominator is of linear order. From the linear approximation, $\phi$, we read off that $\left( \dfrac{\partial f}{\partial x}, \dfrac{\partial f}{\partial y} \right) \Big|_{(-2,1)} = (1, -4)$.  ■

More generally, the linear function $\phi(\boldsymbol{x} - \boldsymbol{x}_0)$ in Definition 2.4 is

$$\phi(\boldsymbol{x} - \boldsymbol{x}_0) = \frac{\partial f}{\partial x_1}\Big|_0 (x_1 - x_{1,0}) + \cdots + \frac{\partial f}{\partial x_n}\Big|_0 (x_n - x_{n,0}).$$

That is, the coefficients $c_1, \ldots, c_n$ of the respective linear factors are simply the $n$ partial derivatives of $f(\boldsymbol{x})$ evaluated at the point $\boldsymbol{x}_0$, which means that $f : \mathbb{R}^n \longrightarrow \mathbb{R}$ is differentiable at $\boldsymbol{x}_0$ if the limit vanishes:

$$\lim_{\boldsymbol{\Delta x} \to 0} \frac{f(x_{0,1} + \Delta x_1, \ldots, x_{0,n} + \Delta x_n) - f(x_{0,1}, \ldots, x_{0,n}) - \sum_{i=1}^{n} \left.\frac{\partial f}{\partial x_i}\right|_0 \Delta x_i}{|\boldsymbol{\Delta x}|} = 0.$$

**Theorem 2.1**
*A function $f : \mathbb{R}^n \longrightarrow \mathbb{R}$ with* **continuous partial derivatives** $\dfrac{\partial f}{\partial x_i}$ *in the neighbourhood of a point $\boldsymbol{x}_0$ is* **differentiable** *at $\boldsymbol{x}_0$.*

This theorem on the continuity of partial derivatives as a condition for differentiability inspires the pictorial interpretation in Figure 2.14.

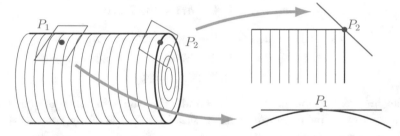

**Figure 2.14** The relation between partial differentiability and differentiability.

At $P_1$ the surface is partially differentiable and differentiable, but at $P_2$ it has limited partial differentiability and so is not differentiable there.

Now for two important theorems (for proofs see [1] or similar texts):

**Theorem 2.2**
*A differentiable function is continuous.*

A continuous function is not necessarily a differentiable function.

**Theorem 2.3**
*A differentiable function is partially differentiable.*

A partially differentiable function is not necessarily a differentiable function.

**Table 2.1:   A pictorial table of differentiable functions (not exhaustive)**

| Function | Diff'ble at $(x_0, y_0)$? | Why or why not? |
|---|:---:|---|
| | *No* | Limit does not exist at points on the red line. Function is not continuous. No tangent plane. |
| | *No* | Limit exists and function is continuous across the red line. But not *all* partial derivatives exist. No tangent plane. |
| | *No* | Function is continuous and all partial derivatives exist, but they are *not continuous* at one point. No tangent plane! (See Mastery Check 2.8.) |
| | **Yes** | Function is continuous and *all* partial derivatives exist and are continuous at $(x_0, y_0)$. There exists a unique tangent plane! |

So, it appears that for $f : \mathbb{R} \longrightarrow \mathbb{R}$, the function is differentiable at a point if the derivative exists, but for $f : \mathbb{R}^2 \longrightarrow \mathbb{R}$, the partial derivatives have to exist *and be continuous* in an open circle about the point. The following Example and Mastery Check make this clear.

■    **Example 2.9:**

Consider the function $f : \mathbb{R}^2 \mapsto \mathbb{R}$, $f(x, y) = x \arctan(y/x)$, $f(0, y) = 0$. We wish to discuss the continuity of $f$, and the existence and continuity of $f_x$ and $f_y$ at points on the $x$-axis.

We have $\lim_{x \to 0} f(x, y) = \lim_{x \to 0} x \arctan(y/x) = 0 = f(0, y)$ for all $y$, since $|\arctan(y/x)| < \pi/2$.

The function is continuous for all points on the $x$-axis.

We have, for $y \neq 0$, $f_x = \arctan(y/x) - \dfrac{xy}{x^2 + y^2}$.

Then for $y > 0$, $\lim_{x \to 0+} f_x = \dfrac{\pi}{2} - 0 = \dfrac{\pi}{2}$, $\lim_{x \to 0-} f_x = -\dfrac{\pi}{2} + 0 = -\dfrac{\pi}{2}$;

and for $y < 0$, $\lim_{x \to 0+} f_x = -\dfrac{\pi}{2} + 0 = -\dfrac{\pi}{2}$, $\lim_{x \to 0-} f_x = \dfrac{\pi}{2} - 0 = \dfrac{\pi}{2}$.

Thus $f_x$ is not continuous along $x = 0$ for $y \neq 0$. Also,

$$f_y = \frac{x^2}{x^2 + y^2}, \qquad \lim_{x \to 0+} f_y = \lim_{x \to 0-} f_y = 0.$$

If we define $f_y(0, 0) = 0$, then $f_y$ exists and is continuous on $x = 0$.                    ■

✍    **Mastery Check 2.8:**

Consider the function $f : \mathbb{R}^2 \longrightarrow \mathbb{R}$ defined by

$$z = f(x, y) = \frac{xy}{\sqrt{x^2 + y^2}} \text{ if } x^2 + y^2 \neq 0, \ f(0, 0) = 0,$$

whose graph appears at the end of this exercise in Figure 2.15. We wish to investigate the behaviour of $f$ near $(0, 0)$.

(1)  Find the two partial derivatives for $(x, y) \neq (0, 0)$.

(2)  Show using Definition 2.3 that both partial derivatives are zero at $(0, 0)$. Thus, the tangent plane at $(0, 0)$, if it exists, *must coincide with the plane $z = 0$.*

(3) But the partial derivatives are not continuous. Show that these limits are not all the same:

$$\lim_{x \to 0}\left(\lim_{y \to 0}\frac{\partial f}{\partial x}\right), \quad \lim_{y \to 0^-}\left(\lim_{x \to 0}\frac{\partial f}{\partial x}\right), \quad \lim_{y \to 0^+}\left(\lim_{x \to 0}\frac{\partial f}{\partial x}\right).$$

(A similar result holds for the other derivative.)

So, we do not expect the tangent plane to exist.

(4) Now recall the properties of a tangent plane as outlined in Definition 2.4. See if you can construct the expression
$$\Delta z = f(\boldsymbol{x}) - \phi(\boldsymbol{x} - \boldsymbol{x}_0) = f(\boldsymbol{x}_0) + |\boldsymbol{\Delta x}|\rho(\boldsymbol{\Delta x}) \quad \text{for the special case}$$
that $\boldsymbol{x}$ lies on the line $y = x$, at distance

$$|\boldsymbol{\Delta x}| = \sqrt{\Delta x^2 + \Delta y^2} \text{ from } \boldsymbol{x}_0 = (0,0), \text{ with } \Delta y = \Delta x.$$

That is, find $\rho(\boldsymbol{\Delta x})$.

Use this result to decide whether a tangent plane exists.

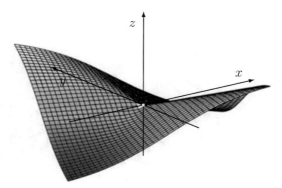

**Figure 2.15**  The graph of $z = \dfrac{xy}{\sqrt{x^2 + y^2}},\ x, y \neq 0.$

# 2.E  Directional derivatives and the gradient

### The directional derivative

Thus far we have established that the partial derivatives of a function $f :$ $\mathbb{R}^2 \longrightarrow \mathbb{R}$ have the properties that:

$\dfrac{\partial f}{\partial x}\Big|_{\substack{\text{some} \\ \text{point}}}$ = the rate of change (that is, the slope) of $f$ in the (positive) $x$-direction at "some point".

$\dfrac{\partial f}{\partial y}\Big|_{\substack{\text{some} \\ \text{point}}}$ = the rate of change (that is, the slope) of $f$ in the (positive) $y$-direction at "some point".

These interpretations now beg the question: What if we wanted to find the rate of change of $f$ in some other direction, such as $\boldsymbol{u}$ depicted in Figure 2.16?

Suppose $f$ is given and we know it is differentiable at a point $(x_0, y_0)$ and we wanted the rate of change of $f$ in that particular direction $\boldsymbol{u}$. We may now combine all the ingredients that go into the limit definition of the derivative in Equation (2.1) on Page 52 and suppose, in addition, that $\boldsymbol{u} = (u, v)$ is a given *unit* vector in the $xy$-plane.

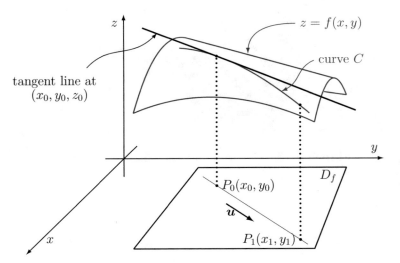

**Figure 2.16** The tangent vector in an arbitrary direction.

We notice first that the two points $P_0$ and $P_1$ appearing in Equation (2.1) define a vector in the $xy$-plane, parallel to our given $\boldsymbol{u}$. Hence,

$$\boldsymbol{\Delta x} \quad = \quad \overrightarrow{P_0 P_1} \quad = \quad \overset{(\quad \Delta x \;,\; \Delta y \quad)}{(x_1 - x_0, y_1 - y_0)} \quad = \quad t \cdot \boldsymbol{u} \quad = \quad t \cdot (u, v).$$

scalar multiplier

unit vector in direction $\overrightarrow{P_0 P_1}$

Now we may re-consider the general expression for the derivative of $f$ in Equation (2.1) which we rewrite for this special case.

---

**Definition 2.5**

*The derivative limit:*

$$\lim_{\boldsymbol{\Delta x} \to 0} \frac{f(x_1, y_1) - f(x_0, y_0)}{|\boldsymbol{\Delta x}|} = \lim_{\boldsymbol{\Delta x} \to 0} \frac{f(x_0 + tu, y_0 + tv) - f(x_0, y_0)}{|t\boldsymbol{u}|}$$

$$= \lim_{t \to 0} \frac{f(x_0 + tu, y_0 + tv) - f(x_0, y_0)}{|t|}, \; (|\boldsymbol{u}|=1),$$

*if it exists, is called* **the directional derivative of $f$ in the direction of $\boldsymbol{u}$ at $(x_0, y_0)$.**

---

Commonly used notations for the directional derivative include

$$D_{\boldsymbol{u}} f(\boldsymbol{x}_0) \text{ and } \frac{\mathrm{d}f}{\mathrm{d}\boldsymbol{u}}(\boldsymbol{x}_0).$$

To calculate the directional derivative, there are two alternatives: Either we use the above definition (which may be necessary if the function is *not continuous or not differentiable*), or defer to the following theorem.

---

**Theorem 2.4**

*If $f : \mathbb{R}^2 \longrightarrow \mathbb{R}$ is a differentiable function, then*

$$D_{\boldsymbol{u}} f(\boldsymbol{x}_0) = \frac{\mathrm{d}f}{\mathrm{d}\boldsymbol{u}}(\boldsymbol{x}_0) = \frac{\partial f}{\partial x}\Big|_0 u + \frac{\partial f}{\partial y}\Big|_0 v.$$

---

For the conditions stated, this theorem is easy to prove.

### ■  Proof:

In light of Definition 2.4, $f$ differentiable means

$$\frac{f(\boldsymbol{x}_0 + t\boldsymbol{u}) - f(\boldsymbol{x}_0)}{t} = \frac{1}{t}\left( f(\boldsymbol{x}_0) + \sum_{i=1}^n \frac{\partial f}{\partial x_i}\Big|_0 tu_i + t\rho(t) - f(\boldsymbol{x}_0)\right)$$

$$= \sum_{i=1}^n \frac{\partial f}{\partial x_i}\Big|_0 u_i + \rho(t) \longrightarrow \sum_{i=1}^n \frac{\partial f}{\partial x_i}\Big|_0 u_i \quad \text{as } t \to 0 \text{ since } \rho(t) \to 0.$$

■

The above simple proof suggests that we can easily extend the definition of a directional derivative and its convenient form to $f : \mathbb{R}^n \longrightarrow \mathbb{R}$. For these functions we have

$$D_{\boldsymbol{u}} f(\boldsymbol{x}_0) = \frac{\partial f}{\partial x_1}\Big|_0 u_1 + \frac{\partial f}{\partial x_2}\Big|_0 u_2 + \cdots + \frac{\partial f}{\partial x_n}\Big|_0 u_n \tag{2.2}$$

where $\boldsymbol{u} = (u_1, u_2, \ldots, u_n)$ and $|\boldsymbol{u}| = 1$.

For a differentiable function what the directional derivative gives us is the slope of the tangent plane in the direction $\boldsymbol{u}$!

### Gradient of a scalar function

Let's look a little more closely at what we use to calculate the directional derivative of a differentiable function. In the general case of $f : \mathbb{R}^n \longrightarrow \mathbb{R}$ we interpret Equation (2.2) as the scalar product of two vectors:

$$D_{\boldsymbol{u}} f(\boldsymbol{x}_0) = \sum_{i=1}^n \frac{\partial f}{\partial x_i}\Big|_0 u_i$$

$$= \underbrace{\left( \frac{\partial f}{\partial x_1}\Big|_0, \frac{\partial f}{\partial x_2}\Big|_0, \ldots, \frac{\partial f}{\partial x_n}\Big|_0 \right)}_{\text{a new vector called } \ldots} \cdot \overbrace{(u_1, u_2, \ldots, u_n)}^{\text{direction of interest}}$$

---

**Definition 2.6**

*The **gradient vector function** of a function $f : \mathbb{R}^n \longrightarrow \mathbb{R}$ at a point $\boldsymbol{x} \in D_f$ is defined as*

$$\mathrm{grad} f(\boldsymbol{x}) = \left( \frac{\partial f}{\partial x_1}, \frac{\partial f}{\partial x_2}, \ldots, \frac{\partial f}{\partial x_n} \right) \equiv \boldsymbol{\nabla} f(\boldsymbol{x})$$

*— we say "grad $f$" or "del $f$".*

Therefore, for  a differentiable $f : \mathbb{R}^n \longrightarrow \mathbb{R}$, the directional derivative of $f(\boldsymbol{x}_0)$ in the direction $\boldsymbol{u}$ is the scalar product of the gradient of $f$ evaluated at $\boldsymbol{x}_0$ and the *unit* vector $\boldsymbol{u}$.

$$D_{\boldsymbol{u}} f(\boldsymbol{x}_0) = \operatorname{grad} f(\boldsymbol{x}_0) \cdot \boldsymbol{u} = \boldsymbol{\nabla} f(\boldsymbol{x}_0) \cdot \boldsymbol{u}.$$

### ■  Example 2.10:

Consider the function $f(x, y, z) = xy^3 + yz^2$. What is the directional derivative at the point $(1, 2, -1)$ in the direction $\boldsymbol{u} = 2\mathbf{e}_1 + \mathbf{e}_2 + 2\mathbf{e}_3$?

The gradient of $f$ is

$$\boldsymbol{\nabla} f = y^3 \mathbf{e}_1 + (3xy^2 + z^2)\mathbf{e}_2 + 2yz\mathbf{e}_3.$$

At $(1, 2, -1)$ this is $\boldsymbol{\nabla} f = 8\mathbf{e}_1 + 13\mathbf{e}_2 - 4\mathbf{e}_3$.

The direction of $\boldsymbol{u} = 2\mathbf{e}_1 + \mathbf{e}_2 + 2\mathbf{e}_3$ is

$$\boldsymbol{n} = \frac{\boldsymbol{u}}{|\boldsymbol{u}|} = \frac{2}{3}\mathbf{e}_1 + \frac{1}{3}\mathbf{e}_2 + \frac{2}{3}\mathbf{e}_3.$$

The directional derivative we require is the scalar product of these:

$$\boldsymbol{\nabla} f \cdot \boldsymbol{n} = (8\mathbf{e}_1 + 13\mathbf{e}_2 - 4\mathbf{e}_3) \cdot \left(\frac{2}{3}\mathbf{e}_1 + \frac{1}{3}\mathbf{e}_2 + \frac{2}{3}\mathbf{e}_3\right) = 7.$$

■

### ✍  Mastery Check 2.9:

What is the unit normal to the surface $xy^2 z^3 = 4$ at the point $(1, 2, -1)$?

✍

### ✍  Mastery Check 2.10:

Calculate the directional derivative of $f(x, y, z) = xy + e^{yz} + z$ in the direction $\boldsymbol{u} = (\alpha, \beta, \gamma)$, where $\alpha^2 + \beta^2 + \gamma^2 = 1$, at the point $(1, 1, 0)$.

When you have found your directional derivative, answer this question:

How should $\alpha, \beta, \gamma$ be chosen so that this derivative should be maximal?

✍

**Remarks** — Some all-important facts about the gradient, $\boldsymbol{\nabla} f$:

* $\boldsymbol{\nabla} f(\boldsymbol{x})$ is the generalization to functions of several variables of $\dfrac{\mathrm{d}g}{\mathrm{d}x}$ for a function $g$ of one variable, $x$.

* In 1D, $\dfrac{\mathrm{d}g}{\mathrm{d}x} = 0$ for all $x \in D_g \implies g(x) = \text{const.}$

  For $f : \mathbb{R}^n \longrightarrow \mathbb{R}$, if

  $$\boldsymbol{\nabla} f = \mathbf{0} \text{ for all } \boldsymbol{x} \text{ in an open set in } D_f,$$

  then $f(\boldsymbol{x})$ is constant in that set.

$$\left.\begin{array}{c}\boldsymbol{\nabla} f(x,y,z) = \mathbf{0} \\ \forall \boldsymbol{x} \in D_f\end{array}\right\} \implies \left.\begin{cases}\dfrac{\partial f}{\partial x} = 0 \Rightarrow f = f(y,z) \\[2mm] \dfrac{\partial f}{\partial y} = 0 \Rightarrow f = f(x,z) \\[2mm] \dfrac{\partial f}{\partial z} = 0 \Rightarrow f = f(x,y)\end{cases}\right\} \implies f(x,y,z) = \text{const.}$$

But, quite often $\boldsymbol{\nabla} f = \mathbf{0}$ at some *isolated* point $\boldsymbol{x}_0$: $f$ is then *not* constant. See Section 3.A for further discussion.

* At a point $\boldsymbol{x} \in D_f$, the differentiable function

$$f : \mathbb{R}^n \longrightarrow \mathbb{R} \left\{\begin{array}{l}\text{increases} \\ \text{decreases}\end{array}\right\} \text{most } \textit{rapidly} \text{ in the direction of } \pm\boldsymbol{\nabla} f(\boldsymbol{x}).$$

The maximum rate of change of $f$ is given by $|\boldsymbol{\nabla} f(\boldsymbol{x})|$.

In Section 3.E we will encounter the gradient again, while later in Chapter 5 the "del" operator appears in other guises.

* If $\boldsymbol{\nabla} f(\boldsymbol{x}) \neq \mathbf{0}$ for a differentiable $f : \mathbb{R}^n \longrightarrow \mathbb{R}$ then $\boldsymbol{\nabla} f(\boldsymbol{x})$ is a vector which is normal to a level set, that is, a $\left\{\begin{array}{l}\text{level curve in } \mathbb{R}^2 \\ \text{level surface in } \mathbb{R}^3\end{array}\right\}$ of $f$.

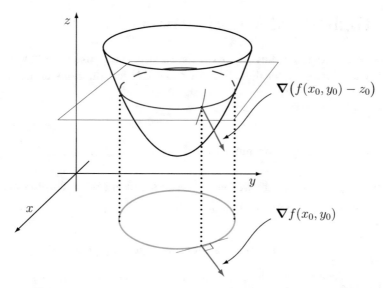

**Figure 2.17** Comparison of the gradient applied in 2D
and 3D circumstances.

In Figure 2.17, it can be seen that $\boldsymbol{\nabla}\big(f(x_0,y_0)-z_0\big)=\dfrac{\partial f}{\partial x}\Big|_0\mathbf{e}_1+\dfrac{\partial f}{\partial y}\Big|_0\mathbf{e}_2-\mathbf{e}_3$
is normal to the 3D surface $f(x,y)-z=$ const. at $(x_0,y_0,z_0)$, while
$\boldsymbol{\nabla}f(x_0,y_0)=\dfrac{\partial f}{\partial x}\Big|_0\mathbf{e}_1+\dfrac{\partial f}{\partial y}\Big|_0\mathbf{e}_2$ is normal to the 2D level curve $f(x,y)=$
const. at $(x_0,y_0)$.

✎   **Mastery Check 2.11:**
Find the equation of the tangent plane to the level surface of $w=f(x,y,z)$
when $w=2$ at the point $(1,1,\pi)$, where $f(x,y,z)=xy\cos(z)+3x$.

✎

✎   **Mastery Check 2.12:**
Find the equation of the tangent plane to the level surface of $w=f(x,y,z)$
at the point $(1,1,\pi)$, where   $f(x,y,z)=xy\cos(z)+3x$.

Hint: We are now working in four dimensions. Consider the level set

$$g(x,y,z,w)=f(x,y,z)-w=0.$$

✎

## 2.F  Higher-order derivatives

By now the reader will have correctly surmised that, just as in the single-variable case, higher-order derivatives are possible for functions of many variables.

Indeed, if

(a) $f : \mathbb{R}^n \longrightarrow \mathbb{R}$ is a continuous function of $x_1, x_2, \ldots, x_n$, and

(b) some given partial derivative $\dfrac{\partial f}{\partial x_i}$ exists and is itself a continuous (not necessarily differentiable) function of $x_1, x_2, \ldots, x_n$,

then $\dfrac{\partial f}{\partial x_i}$ can itself be considered a function of $\boldsymbol{x}$ (independent of $f$).

For a convenient explanation, we shall refer to this particular derivative as $g(\boldsymbol{x}) \left( \equiv \dfrac{\partial f}{\partial x_i}(\boldsymbol{x}) \right)$. We can now consider the partial derivatives of $g$ just as we had done with $f$:

---

**Definition 2.7**

If $\dfrac{\partial g}{\partial x_j} \equiv \lim\limits_{h \to 0} \dfrac{g(x_1, \ldots, x_j + h, \ldots, x_n) - g(x_1, \ldots, x_j, \ldots, x_n)}{h}$ *exists it is called* **a second-order partial derivative of** $f$. *More specifically, it is a second-order mixed partial derivative of* $f$ *w.r.t.* $x_i$ *and* $x_j$.

---

In terms of the original $f$ we have that

$$\frac{\partial g}{\partial x_j} = \frac{\partial}{\partial x_j}\left(\frac{\partial f}{\partial x_i}\right) \equiv \frac{\partial^2 f}{\partial x_j \partial x_i} \qquad \text{— this is the notation used}$$
$$\text{(mostly) in this book}$$

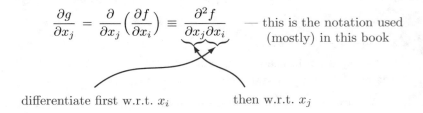

differentiate first w.r.t. $x_i$                     then w.r.t. $x_j$

Note that other notations for the second derivative of $f(x, y)$ are in common use such as

$$\frac{\partial^2 f}{\partial y \partial x} = f_{xy} = f_{12},$$

each of which describes a second-order partial derivative. First, a partial derivative w.r.t. $x$, then a partial derivative w.r.t. $y$. The reader should exercise some care in interpreting the different notations.

We are now implored to explain what higher partial derivatives are. It suffices to consider a function of two variables, $f(x,y)$. If $\dfrac{\partial f}{\partial x}\bigg|_{(x_0,y_0)}$ is the slope of the tangent to $f$ at $(x_0, y_0)$ in the direction of $x$, then, just as in the single-variable case, $\dfrac{\partial^2 f}{\partial x^2}\bigg|_{(x_0,y_0)}$ is the rate of change of the slope in this same direction. It is therefore a measure of the curvature of $f$ in this direction. On the other hand, $\dfrac{\partial^2 f}{\partial y \partial x}\bigg|_{(x_0,y_0)}$ is the rate of change of the $x$-directional slope in the $y$-direction.

A convenient and useful result for so-called *smooth* functions which, apart from their applications in applied contexts (Chapters 3 and 5), relieves some of the stress of interpreting notation, is the following.

---

**Theorem 2.5**

*Suppose $f : \mathbb{R}^n \longrightarrow \mathbb{R}$ is continuous and $\dfrac{\partial f}{\partial x_i}$, $i = 1, 2, \ldots, n$ exist and are continuous in $S_r(\boldsymbol{x}) \subset D_f$ and that both $\dfrac{\partial^2 f}{\partial x_i \partial x_j}$ and $\dfrac{\partial^2 f}{\partial x_j \partial x_i}$ exist and are continuous at $\boldsymbol{x} \in D_f$. Then $\dfrac{\partial^2 f}{\partial x_i \partial x_j} = \dfrac{\partial^2 f}{\partial x_j \partial x_i}$ at $\boldsymbol{x} \in D_f$.*

---

(For the standard proof, see a standard text book such as [1] or [2].)

Note the conditions of the above theorem highlighted in Figure 2.18.

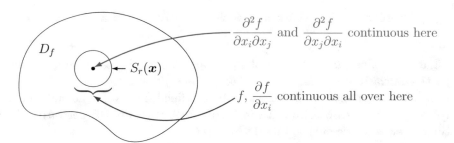

$\dfrac{\partial^2 f}{\partial x_i \partial x_j}$ and $\dfrac{\partial^2 f}{\partial x_j \partial x_i}$ continuous here

$f, \dfrac{\partial f}{\partial x_i}$ continuous all over here

**Figure 2.18**   Conditions for the equivalence of mixed partial derivatives.

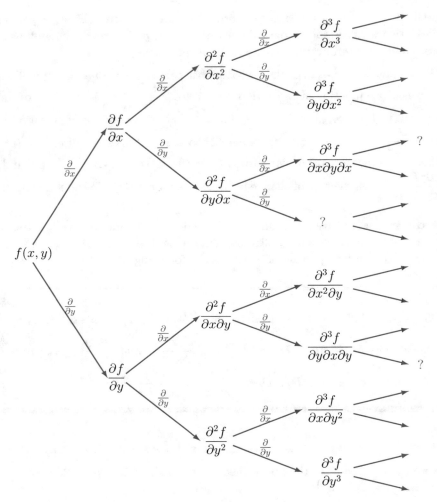

**Figure 2.19**  A chart of possible derivatives of $f : \mathbb{R}^2 \longrightarrow \mathbb{R}$.

Figure 2.19 gives an indication of the scope of possibilities of higher-order partial derivatives for a function (*of two variables*) that is sufficiently differentiable. What constitutes "sufficient" will be defined shortly. In the meantime, referring to the arrow convention in Figure 2.19, what are the derivatives in the positions where question marks appear?

---

**Definition 2.8**

*A function* $f : \mathbb{R}^n \longrightarrow \mathbb{R}$ *with* **continuous derivatives** *up to and including order* $m$ $(0,1,2,\ldots,m)$ *in an open subset of* $D_f$ *is said to be of class* $C^m$.

---

A $C^2$-function thus satisfies $\dfrac{\partial^2 f}{\partial x_i \partial x_j} = \dfrac{\partial^2 f}{\partial x_j \partial x_i}$.

✍   **Mastery Check 2.13:**

Verify that for $f(x,y) = 2 - x^2 y - xy^2$, the mixed derivatives $\dfrac{\partial^2 f}{\partial x \partial y}$ and $\dfrac{\partial^2 f}{\partial y \partial x}$ are equal. Draw the graph of $f(x,y)$ using MATLAB®.

✍

✍   **Mastery Check 2.14:**

Determine all partial derivatives of order 2 of $f(x,y) = \arctan\left(\dfrac{x}{y}\right)$.

Specify any points where $f$ or its derivatives are not defined.

✍

✍   **Mastery Check 2.15:**

Determine all $C^2$ functions $z = f(x,y)$ which satisfy the conditions

$$\frac{\partial z}{\partial x} = xye^x + 1, \quad \frac{\partial z}{\partial y} = (x-1)e^x + 1.$$

Hint: First, check to see whether there exist such functions. Then, find possible antiderivatives to the two conditioning equations.

✍

✍   **Mastery Check 2.16:**

Determine all $C^2$ functions $f(x,y)$ such that

a) $\dfrac{\partial f}{\partial x} = 2x \sin x^2, \quad \dfrac{\partial f}{\partial y} = \cos y.$

b) $\dfrac{\partial f}{\partial x} = 2x + y, \quad \dfrac{\partial f}{\partial y} = 2y + x.$

c) $\dfrac{\partial f}{\partial x} = x + 3yx^2, \quad \dfrac{\partial f}{\partial y} = x^3 + xy.$

✍

# 2.G   Composite functions and the chain rule

We now come to a topic which many find challenging. However, it is so important in multivariable calculus as well as in practice that we will devote some considerable space to it.

What are *composite functions*? These are functions *of functions*, of one or more independent variables. The relationships between the functions and their variable dependencies can be readily represented by *ball-and-stick diagrams* (see below). Although we will retain function names in our analyses, the ball-and-stick diagrams show the relationships between dependent and independent variables. However, the reader should bear in mind that the functions themselves actually provide the links between the variables. These links are illustrated with the help of *domain-and-range diagrams*, which seek to aid understanding not only of the dependencies but also of the conditions that must be satisfied for the composite functions to be defined.

What then is the *chain rule*? The simple truth that this is the process by which one differentiates composite functions is rather unhelpful at this point. It will be necessary to go through the various cases we will be considering, in order of complexity, for this statement to have meaning.

As for illustrating the chain rule — as distinct from visualizing composite functions — we take advantage of the notion that derivatives describe rates of change and so imagine a derivative to represent the regulated flow of water through a sluice gate or *floodgate* from one water reservoir to another. The chain rule, which involves sums of products of derivatives, we shall represent by *floodgate diagrams*: an arrangement of reservoirs fitted with floodgates to regulate water flow. The net flow of water out through the final gate will depend on which gates above it are open (the variables), by how much (the partial derivatives), how two or more flow rates reinforce (the products), and in what combinations (the sums).

In this context, probably more than any other, it is important to distinguish between the independent variable that is involved in a partial derivative and others that are held fixed. To this end we will use notation such as

$$\left(\frac{\partial f}{\partial x}\right)_y \quad \text{and} \quad \left(\frac{\partial F}{\partial u}\right)_v$$

to refer to partial derivatives (here w.r.t. $x$ and $u$, respectively) and the independent variables that are kept constant (here $y$ and $v$, respectively).

## Case 1

This is the simplest example which the student would have encountered in their single-variable calculus course. It nevertheless exhibits all the features inherent in the more complicated multivariable cases to follow. Accordingly, the format we follow in this discussion is repeated in the latter cases. Within this format we itemize the variable dependence of the functions involved including their domains and ranges, the composite function and its domain and range, and finally the appropriate chain rule for derivatives of the composite function.

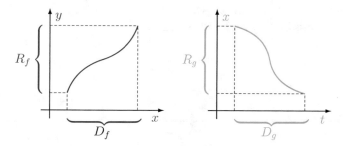

**Figure 2.20**   The graphs of $y = f(x)$ and $x = g(t)$.

Consider two functions $f, g \in C^1(\mathbb{R})$ of one variable.

$$f : \mathbb{R} \longrightarrow \mathbb{R} \qquad\qquad g : \mathbb{R} \longrightarrow \mathbb{R}$$
$$x \longmapsto y = f(x), \qquad\qquad t \longmapsto x = g(t).$$

The domains and ranges of these are shown in Figure 2.20 above. From these functions we form the composite function of the independent variable $t$:

$$y = F(t) = (f \circ g)(t) = f(g(t)).$$

The composite function may be represented schematically by the ball-and-stick diagram in Figure 2.21. The diagram (as with the more complex ones to follow) indicates that the variable $y$ depends on $x$, which in turn depends on $t$. Thus, a variation in $t$ leads to a variation in $x$, which leads to a variation in $y$.

**Figure 2.21**  Ball-and-stick model for $f\big(g(t)\big)$.

The domain and range of $F$, which are based on the sets $D_g$, $R_g$, $D_f$, and $R_f$, must be such that $F$ makes sense. Examine Figure 2.22, from left to right, noting the termini (start and end points) of the arrows.

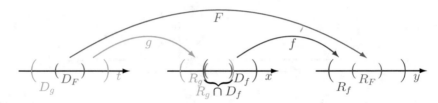

**Figure 2.22**  Conditional definition of $D_F$ and $R_F$.

From the figure follow two set relations and one critical set constraint: the domain of $F$ is a subset of the domain of $g$, the range of $F$ is a subset of the range of $f$, and then there must be a non-empty intersection of the range of $g$ and domain of $f$. In set notation these are summarized as follows.

$$\left.\begin{array}{l} \text{Domain of } F : D_F \subseteq D_g \\ \text{Range of } F : R_F \subseteq R_f \end{array}\right\} \text{ An important condition: } R_g \cap D_f \neq \emptyset.$$

The derivative of $F$ w.r.t. $t$ is given by the chain rule and can be represented schematically by the *floodgate* diagram in Figure 2.23.

$$\frac{\mathrm{d}F}{\mathrm{d}t} = \frac{\mathrm{d}}{\mathrm{d}t}\big(f \circ g\big)(t)$$

$$= \frac{\mathrm{d}f}{\mathrm{d}x}\bigg|_{x=g(t)} \cdot \frac{\mathrm{d}g}{\mathrm{d}t}\bigg|_t$$

$$= \left\{ \begin{array}{c} \text{derivative} \\ \text{of } f \\ \text{evaluated} \\ \text{at } x = g(t) \end{array} \right\} \times \left\{ \begin{array}{c} \text{derivative} \\ \text{of } g \\ \text{evaluated} \\ \text{at } t \end{array} \right\}$$

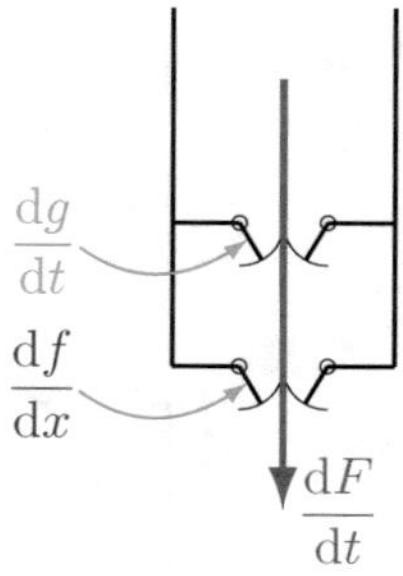

**Figure 2.23**  The floodgate diagram for $\mathrm{d}F/\mathrm{d}t$.

In constructing the relevant floodgate diagram for any given case we aim to compile and represent all the term- and factor-wise contributions to a chain rule derivative. The objective is the net outflow at the bottom. In this case the final flow rate out through the bottom gate of the reservoir $\left(\dfrac{\mathrm{d}F}{\mathrm{d}t}\right)$ is dependent not only on the flow rate from the $x$-reservoir to the $y$-reservoir $\left(\dfrac{\mathrm{d}f}{\mathrm{d}x}\right)$ but also on the flow rate from the $t$-reservoir to the $x$-reservoir $\left(\dfrac{\mathrm{d}g}{\mathrm{d}t}\right)$: one reinforces (multiplies) the other.

A final note concerns the notation used to represent and describe the chain rule derivative. The more commonly seen notation is

$$\frac{\mathrm{d}y}{\mathrm{d}t} = \frac{\mathrm{d}y}{\mathrm{d}x} \cdot \frac{\mathrm{d}x}{\mathrm{d}t}.$$

Although it is intuitive and appealing to express the chain rule in this way (admittedly it is convenient sometimes), this notation can be problematic in some cases.

**Case 2**

We now move on to more complicated functional arrangements. As was remarked earlier, the format for the discussion here remains the same as in Case 1.

Consider function $f$ as before, but now suppose that $g$ is a function of two variables.

$$f : \mathbb{R} \longrightarrow \mathbb{R} \qquad\qquad g : \mathbb{R}^2 \longrightarrow \mathbb{R}$$
$$x \longmapsto y = f(x) \qquad\qquad (s,t) \longmapsto x = g(s,t)$$

The respective domains and ranges are shown schematically in Figure 2.24.

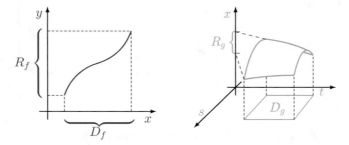

**Figure 2.24** The graphs of $y = f(x)$ and $x = g(s,t)$.

The composite function of the two independent variables, $s$ and $t$, is

$$y = F(s,t) = (f \circ g)(s,t) = f\big(g(s,t)\big).$$

The composite function can here too be represented schematically by a ball-and-stick diagram, but this time a *branch* diagram (Figure 2.25). The variable $y$ depends on $x$, which depends on $s$ and $t$. Consequently, a variation in $s$ or $t$ leads to a variation in $x$, which leads in turn to a variation in $y$.

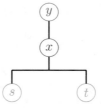

**Figure 2.25** Ball-and-stick for $f\big(g(s,t)\big)$.

The domain and range of $F$, which again are determined by $D_g$, $R_g$, $D_f$, and $R_f$, are such that $F$ makes sense (Figure 2.26).

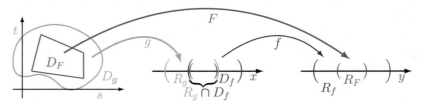

**Figure 2.26** Conditional definition of $D_F$ and $R_F$.

We once again find two set relations and the same set constraint.

$$\left.\begin{array}{l}\text{Domain of } F: D_F \subseteq D_g \\ \text{Range of } F: R_F \subseteq R_f\end{array}\right\} \text{ An important condition: } R_g \cap D_f \neq \emptyset.$$

That is, the domain of $F$ is not bigger than, and can be smaller than, the domain of $g$, and the range of $F$ is not bigger and can be smaller than the range of $f$. All is dictated by the size of $R_g \cap D_f$.

The derivatives of $F$ w.r.t. $s$ and $t$ are now *partial derivatives* given by the chain rule, and represented by their respective *floodgate diagrams*: Figures 2.27 and 2.28.

$$\left(\frac{\partial F}{\partial s}\right)_t = \frac{\partial}{\partial s}(f \circ g)(s,t)$$
$$t \text{ — held fixed}$$
$$= \left.\frac{\mathrm{d}f}{\mathrm{d}x}\right|_{x=g(s,t)} \cdot \left(\frac{\partial g}{\partial s}\right)_t$$

$$\left\{\begin{array}{c}\text{full} \\ \text{derivative} \\ \text{of } f \\ \text{evaluated} \\ \text{at } x = g(s,t)\end{array}\right\} \times \left\{\begin{array}{c}\text{partial} \\ \text{derivative} \\ \text{of } g \\ \text{evaluated} \\ \text{at } (s,t)\end{array}\right\}$$

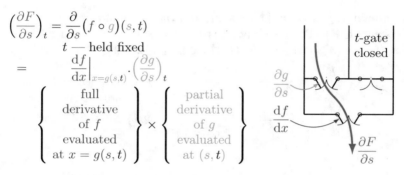

**Figure 2.27**   The floodgate diagram for $\partial F/\partial s$.

$$\left(\frac{\partial F}{\partial t}\right)_s = \frac{\partial}{\partial t}(f \circ g)(s,t)$$
$$s \text{ — held fixed}$$
$$= \left.\frac{\mathrm{d}f}{\mathrm{d}x}\right|_{x=g(s,t)} \cdot \left(\frac{\partial g}{\partial t}\right)_s$$

$$\left\{\begin{array}{c}\text{full} \\ \text{derivative} \\ \text{of } f \\ \text{evaluated} \\ \text{at } x = g(s,t)\end{array}\right\} \times \left\{\begin{array}{c}\text{partial} \\ \text{derivative} \\ \text{of } g \\ \text{evaluated} \\ \text{at } (s,t)\end{array}\right\}$$

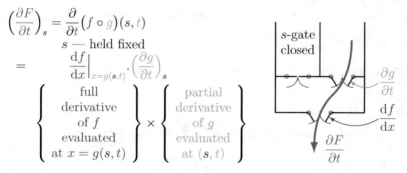

**Figure 2.28**   The floodgate diagram for $\partial F/\partial t$.

In this case (in contrast with Case 1) the partial derivatives of $F$ mean that one variable is kept fixed, and thus its associated gate remains closed, giving no contribution to the flow out through the bottom floodgate.

We demonstrate how this case works with an example and leave the reader with an exercise to consolidate their understanding.

■ **Example 2.11:**

Consider the functions $f : x \longrightarrow y = f(x)$, and $g : (s,t) \longrightarrow x = g(s,t)$. We wish to find the domain $D_F$ of the composite function $F : (s,t) \longrightarrow y = (f \circ g)(s,t)$, and the derivatives $\left(\dfrac{\partial F}{\partial s}\right)_t$, $\left(\dfrac{\partial F}{\partial t}\right)_s$, when $f(x) = \ln x$, $g(s,t) = s(1 - t^2)$.

The domain of $f$ is $D_f = \{x : x > 0\}$, and the range of $g$ is $R_g = \{x : x \in \mathbb{R}\}$. The intersection is $\{x : x = s(1 - t^2) > 0\}$, that is, $D_F = \{(s,t) : (s > 0 \text{ and } |t| < 1) \cup (s < 0 \text{ and } |t| > 1)\}$.

$$\left(\frac{\partial F}{\partial s}\right)_t = \frac{df}{dx}\frac{\partial g}{\partial s} = \frac{1}{x}\bigg|_{x=s(1-t^2)} (1 - t^2) = \frac{1}{s}.$$

$$\left(\frac{\partial F}{\partial t}\right)_s = \frac{df}{dx}\frac{\partial g}{\partial t} = \frac{1}{x}\bigg|_{x=s(1-t^2)} (-2st) = -\frac{2t}{1 - t^2}.$$

■

✎ **Mastery Check 2.17:**

Consider the function: $y = f(x) = \arcsin x$, where $x = g(s,t) = s^2 + \dfrac{1}{t}$. What are $D_f$, $R_f$, $D_g$, and $R_g$?

Determine where $y = F(s,t)$ makes sense, and then find (if possible) the partial derivatives $\left(\dfrac{\partial F}{\partial s}\right)_t$ and $\left(\dfrac{\partial F}{\partial t}\right)_s$.

Note: the final result should be expressed in terms of $s$ and $t$!          ✎

**Case 3**

Consider two functions $f, g \in C^1(\mathbb{R}^2)$ of two variables.

$$f : \mathbb{R}^2 \longrightarrow \mathbb{R} \qquad\qquad g : \mathbb{R}^2 \longrightarrow \mathbb{R}^2$$
$$(x, y) \longmapsto z = f(x, y) \qquad\qquad (s, t) \longmapsto (x = g_1(s,t), y = g_2(s,t))$$

The composite function $F$ of two variables $s$ and $t$ derived from $f$ and $g$ is

$$z = F(s,t) = (f \circ g)(s,t) = f(g_1(s,t), g_2(s,t)).$$

This composite function is represented by the more elaborate branch model of dependent and independent variables shown in Figure 2.29.

This time $z$ depends on $x$ and $y$, and both $x$ and $y$ depend on $s$ and $t$.

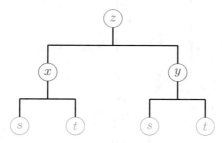

**Figure 2.29**  Ball-and-stick for $f\big(g_1(s,t), g_2(s,t)\big)$.

The domain and range of $F$, which are dictated by $D_f$ and $R_f$, and $D_g$ and $R_g$, are such that $F$ makes sense. In analogy with Cases 1 and 2, the same set conditions and set constraint can be established from the depiction in Figure 2.30:

$$\left.\begin{array}{l} \text{Domain of } F : D_F \subseteq D_g \\ \text{Range of } F : R_F \subseteq R_f \end{array}\right\} \text{ An important condition: } R_g \cap D_f \neq \emptyset.$$

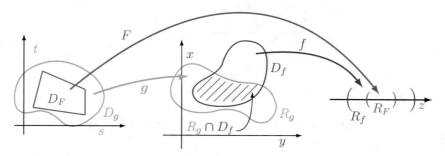

**Figure 2.30**  Conditional definition of $D_F$ and $R_F$.

The partial derivatives of $F$ w.r.t. $s$ and $t$ are given by the chain rule:

$$\left(\frac{\partial F}{\partial s}\right)_t = \left(\frac{\partial f}{\partial x}\right)_y\Bigg|_{\substack{x=g_1(s,t)\\y=g_2(s,t)}} \cdot \left(\frac{\partial g_1}{\partial s}\right)_t + \left(\frac{\partial f}{\partial y}\right)_x\Bigg|_{\substack{x=g_1(s,t)\\y=g_2(s,t)}} \cdot \left(\frac{\partial g_2}{\partial s}\right)_t$$

$t$ — held constant

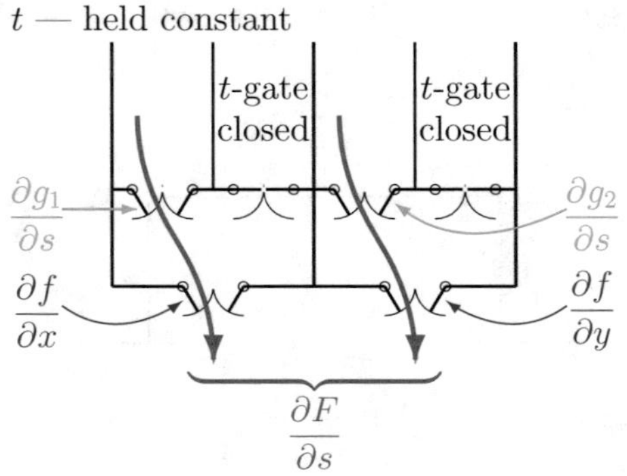

**Figure 2.31** The floodgate diagram for $\partial F/\partial s$.

$$\left(\frac{\partial F}{\partial t}\right)_s = \left(\frac{\partial f}{\partial x}\right)_y\Bigg|_{\substack{x=g_1(s,t)\\y=g_2(s,t)}} \cdot \left(\frac{\partial g_1}{\partial t}\right)_s + \left(\frac{\partial f}{\partial y}\right)_x\Bigg|_{\substack{x=g_1(s,t)\\y=g_2(s,t)}} \cdot \left(\frac{\partial g_2}{\partial t}\right)_s$$

$s$ — held constant

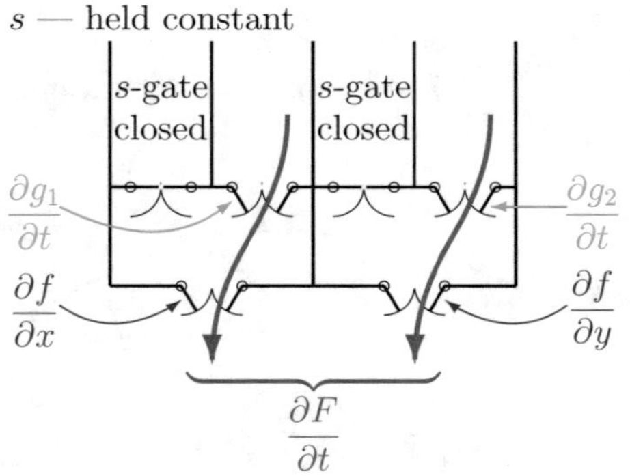

**Figure 2.32** The floodgate diagram for $\partial F/\partial t$.

In each of the cases shown in Figures 2.31 and 2.32, both bottom floodgates $\left(\dfrac{\partial f}{\partial x} \text{ and } \dfrac{\partial f}{\partial y}\right)$ are open and contribute to the total flow, but with strengths that are modulated by the floodgates above them, that is, by the partial derivatives of $g_1$ and $g_2$.

### ✍  Mastery Check 2.18:

Consider the following function:

$$z = f(x,y) = \sin(x^2 y), \text{ where } x = g_1(s,t) = st^2 \text{ and } y = g_2(s,t) = s^2 + \frac{1}{t}.$$

Let $F(s,t) = (f \circ \boldsymbol{g})(s,t)$. Find $\left(\dfrac{\partial F}{\partial s}\right)_t$ and $\left(\dfrac{\partial F}{\partial t}\right)_s$, if they make sense.

✍

### Case 4

As a final example, consider two functions $f \in C^1(\mathbb{R}^3)$, $\boldsymbol{g} \in C^1(\mathbb{R}^2)$. This is a mixed case where the "outer" function, $f$, depends on an independent variable both directly and indirectly.

$$f : \mathbb{R}^3 \longrightarrow \mathbb{R} \qquad\qquad\qquad \boldsymbol{g} : \mathbb{R}^2 \longrightarrow \mathbb{R}^2$$
$$(x,y,t) \longmapsto z = f(x,y,t) \qquad\qquad (s,t) \longmapsto (x = g_1(s,t), y = g_2(s,t))$$

The composite function $F$ of two variables $s$ and $t$ formed from $f$ and $\boldsymbol{g}$ is

$$z = F(s,t) = f\big(g_1(s,t), g_2(s,t), t\big).$$

The ball-and-stick branch model appropriate for this example is shown below in Figure 2.33.

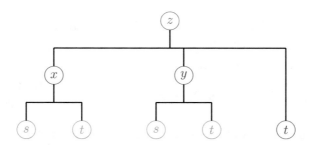

**Figure 2.33**  The complex ball-and-stick diagram
for $f\big(g_1(s,t), g_2(s,t), t\big)$.

As in all previous examples, the sets $D_f$, $R_f$, $D_g$, and $R_g$ establish the domain and range of $F$ so that $F$ makes sense. However, this time there is the added complication of the appearance of a common independent variable, $t$.

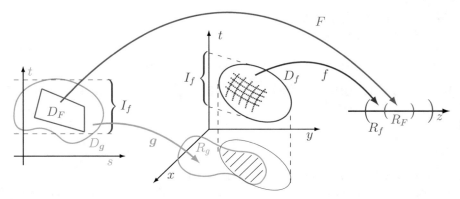

**Figure 2.34** Conditional definition of $D_F$ and $R_F$.

Note the complex intersections of domains and ranges in Figure 2.34. The particular complication here is the fact that $t$-values must lie in $I_f$, the $t$-interval making up *one* dimension of $D_f$, as well as in the *one* dimension within the domain of $g$.

To reconcile the different sets, let $PD_f$ be the projection of $D_f$ in the $xy$-plane and consider the infinite strip $I_f \times \mathbb{R} \subset \mathbb{R}^2$ in the $st$-plane. The rectangular strip $I_f \times \mathbb{R}$ is shown in the left-hand diagram in Figure 2.34.

We can now formally establish the range and domain of $F$:

$$\left.\begin{array}{l} \text{Domain of F: } D_F \subseteq D_g \cap (I_f \times \mathbb{R}) \\ \text{Range of F: } R_F \subseteq R_f \end{array}\right\}.$$

$$\text{There are two important conditions: } \begin{cases} R_g \cap PD_f \neq \emptyset \\ D_g \cap (I_f \times \mathbb{R}) \neq \emptyset. \end{cases}$$

The preceding two conditions are critically important for the validity of the composite function. The domain of $F$ must be consistent with $R_g \cap PD_f$, but the allowed $t$-values in $D_g$ must also be within $I_f$. Consequently, $t$ in $f$ is *not* independent of $t$ in $g$! Note also that the limits of the interval $I_f$ *may* depend on $x$ and $y$ values in $PD_f$.

We will meet this idea again in another context in Section 4.G.

The partial derivative of $F$ w.r.t. $s$ is given by the chain rule, largely identical to the previous case:

$$\left(\frac{\partial F}{\partial s}\right)_t = \left(\frac{\partial f}{\partial x}\right)_{y,t}\bigg|_{\substack{x=g_1(s,t)\\y=g_2(s,t)}} \cdot \left(\frac{\partial g_1}{\partial s}\right)_t + \left(\frac{\partial f}{\partial y}\right)_{x,t}\bigg|_{\substack{x=g_1(s,t)\\y=g_2(s,t)}} \cdot \left(\frac{\partial g_2}{\partial s}\right)_t$$

$t$ — held constant

The corresponding floodgate model for $\left(\dfrac{\partial F}{\partial s}\right)_t$ is also effectively as appears in the preceding case. On the other hand, the partial derivative of $F$ w.r.t. $t$ is given by a version of the chain rule that has three contributions (see Figure 2.35).

$$\left(\frac{\partial F}{\partial t}\right)_s = \left(\frac{\partial f}{\partial x}\right)_{y,t}\bigg|_{\substack{x=g_1(s,t)\\y=g_2(s,t)}} \cdot \left(\frac{\partial g_1}{\partial t}\right)_s + \left(\frac{\partial f}{\partial y}\right)_{x,t}\bigg|_{\substack{x=g_1(s,t)\\y=g_2(s,t)}} \cdot \left(\frac{\partial g_2}{\partial t}\right)_s + \left(\frac{\partial f}{\partial t}\right)_{x,y}$$

$s$ — held constant

total variation          indirect variation          direct variation

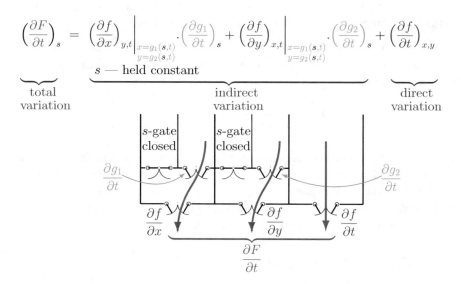

**Figure 2.35**  The floodgate diagram for $\partial F/\partial t$.

The indirect contribution to $\left(\dfrac{\partial F}{\partial t}\right)_s$ is as in the preceding case, but now there is an extra contribution from the direct dependence on $t$. This is reflected in Figure 2.35 by the feature of the reservoir above the floodgate $\left(\dfrac{\partial f}{\partial t}\right)_{x,y}$ having no other influences, while there are $t$-reservoirs that influence (multiply or reinforce) the other contributions.

The reader should compare the expression for the partial derivative $\left(\dfrac{\partial F}{\partial t}\right)_s$ with the expression one would write using the less precise notation that treats $x, y, z$, as both dependent and independent variables. With $z = F(s, t)$ and $z = f(x, y, t)$, the chain rule for the partial derivative with respect to $t$ would then be written

$$\frac{\partial z}{\partial t} = \underbrace{\frac{\partial z}{\partial x} \cdot \frac{\partial x}{\partial t} + \frac{\partial z}{\partial y} \cdot \frac{\partial y}{\partial t}}_{} + \underbrace{\frac{\partial z}{\partial t}}_{}.$$

$$\underbrace{\phantom{\frac{\partial z}{\partial t}}}_{\substack{\text{total} \\ \text{variation}}} \quad \underbrace{\phantom{\frac{\partial z}{\partial x}}}_{\substack{\text{indirect} \\ \text{variation}}} \quad \underbrace{\phantom{\frac{\partial z}{\partial t}}}_{\substack{\text{direct} \\ \text{variation}}}$$

We see that by being imprecise we arrive at an expression involving two terms with the same notation but which mean different things!

If the reader insists on using $x, y, z$ as both functions and independent variables instead of just as independent variables, then they should at least write the derivatives as

$$\frac{\partial z}{\partial t} = \left(\frac{\partial z}{\partial x}\right)_{y,t} \cdot \frac{\partial x}{\partial t} + \left(\frac{\partial z}{\partial y}\right)_{x,t} \cdot \frac{\partial y}{\partial t} + \left(\frac{\partial z}{\partial t}\right)_{x,y}.$$

We have completed our coverage of composite functions and their first partial derivatives. Of course, these four cases do not exhaust all possibilities. However, the reader may well discover that more complicated cases of composite functions and their respective partial derivatives may be readily if not easily constructed by generalizing the exposition given here.

## ✎ Mastery Check 2.19:

Consider the function

$$z = f(x, y, t) = xt \cos y,$$

where

$$x = g_1(s, t) = st + 1, \quad y = g_2(s, t) = s^2 + t^2.$$

Let $\left(D_f\right)_t \subseteq \mathbb{R}^2$ denote the domain of $f$ when $t$ is held constant. Your task is first to establish that $\left(D_f\right)_t \cap R_g \neq \emptyset$, and then to determine the partial derivatives of $F(s, t) = \left(f \circ g\right)(s, t)$ w.r.t. $s$ and $t$.

✎

**Second derivatives and the chain rule**

Applications involving the chain rule are not limited to first partial derivatives. So, while it is essential to understand the chain rule in principle, a Master Class in the practical use of the chain rule cannot be complete without a discussion of higher-order partial derivatives. In the author's experience, this is what most students find challenging.

By way of demonstration consider Case 3 again:

$$f : \mathbb{R}^2 \longrightarrow \mathbb{R}, \text{ and } \boldsymbol{g} : \mathbb{R}^2 \longrightarrow \mathbb{R}^2, \text{ with}$$

$$F(s,t) = (f \circ g)(s,t) = f\big(g_1(s,t), g_2(s,t)\big).$$

The first partial derivative of $F$ w.r.t. $s$ is (as on Page 91):

$$\frac{\partial F}{\partial s} = \frac{\partial f}{\partial x} \cdot \frac{\partial g_1}{\partial s} + \frac{\partial f}{\partial y} \cdot \frac{\partial g_2}{\partial s},$$

where for convenience we have suppressed parentheses and their subscripts.

If we now want a second derivative, say $\dfrac{\partial^2 F}{\partial t \, \partial s}$, then we must take note of two facts:

(a) $\dfrac{\partial F}{\partial s}$ is the sum of *products* of functions!

(b) $\dfrac{\partial f}{\partial x}$ and $\dfrac{\partial f}{\partial y}$ are *two new composite functions of s and t!* Let's denote these by $K$ and $H$.

Hence, using

$$\frac{\partial f}{\partial x} = k\big(x(s,t), y(s,t)\big) = K(s,t), \quad \frac{\partial f}{\partial y} = h\big(x(s,t), y(s,t)\big) = H(s,t),$$

the first partial derivative of $F$ w.r.t. $s$ will become

$$\frac{\partial F}{\partial s} = K(s,t) \cdot \frac{\partial g_1}{\partial s} + H(s,t) \cdot \frac{\partial g_2}{\partial s}.$$

In this guise, the first partial derivative is more clearly seen to be a sum of products of functions of $s$ and $t$. Consequently, in taking a second partial derivative — any second partial derivative for that matter — we must take the following steps in the order given:

**Step 1:** use the product rule of differentiation;

**Step 2:** use the chain rule *again*, this time on $K$ and $H$;

**Step 3:** express everything in terms of the independent variables of $F$.

As we said, these actions *must* be taken *in this order!*

**Step 1:** Differentiate the sum of products using the product rule

$$\frac{\partial^2 F}{\partial t\,\partial s} = \frac{\partial}{\partial t}\left(\frac{\partial F}{\partial s}\right) = \frac{\partial}{\partial t}\left(K(s,t)\frac{\partial g_1}{\partial s}\right) + \frac{\partial}{\partial t}\left(H(s,t)\frac{\partial g_2}{\partial s}\right)$$

$$= K(s,t)\frac{\partial^2 g_1}{\partial t\,\partial s} + H(s,t)\frac{\partial^2 g_2}{\partial t\,\partial s} + \frac{\partial K(s,t)}{\partial t}\cdot\frac{\partial g_1}{\partial s} + \frac{\partial H(s,t)}{\partial t}\cdot\frac{\partial g_2}{\partial s}$$

product rule                  product rule

4 terms

The first two terms are finished.

**Step 2:** After the product rule, apply the *chain* rule (again) on the second two terms:

$$\frac{\partial K}{\partial t} = \frac{\partial k}{\partial x}\cdot\frac{\partial g_1}{\partial t} + \frac{\partial k}{\partial y}\cdot\frac{\partial g_2}{\partial t} = \frac{\partial^2 f}{\partial x^2}\cdot\frac{\partial g_1}{\partial t} + \frac{\partial^2 f}{\partial y\,\partial x}\cdot\frac{\partial g_2}{\partial t}.$$

$$\frac{\partial H}{\partial t} = \frac{\partial h}{\partial x}\cdot\frac{\partial g_1}{\partial t} + \frac{\partial h}{\partial y}\cdot\frac{\partial g_2}{\partial t} = \frac{\partial^2 f}{\partial x\,\partial y}\cdot\frac{\partial g_1}{\partial t} + \frac{\partial^2 f}{\partial y^2}\cdot\frac{\partial g_2}{\partial t}.$$

**Step 3:** Finally, replace all the $K$ and $H$ factors with the $f$, $g_1$, and $g_2$ factors (with all derivatives of $f$ evaluated at $x = g_1(s,t)$ and $y = g_2(s,t)$):

$$\frac{\partial^2 F}{\partial t\,\partial s} = \frac{\partial f}{\partial x}\bigg|_{\substack{x=g_1\\y=g_2}}\frac{\partial^2 g_1}{\partial t\,\partial s} + \frac{\partial f}{\partial y}\bigg|_{\substack{x=g_1\\y=g_2}}\frac{\partial^2 g_2}{\partial t\,\partial s}$$

$$+ \frac{\partial g_1}{\partial s}\left(\frac{\partial^2 f}{\partial x^2}\bigg|_{\substack{x=g_1\\y=g_2}}\frac{\partial g_1}{\partial t} + \frac{\partial^2 f}{\partial y\,\partial x}\bigg|_{\substack{x=g_1\\y=g_2}}\frac{\partial g_2}{\partial t}\right)$$

$$+ \frac{\partial g_2}{\partial s}\left(\frac{\partial^2 f}{\partial x\,\partial y}\bigg|_{\substack{x=g_1\\y=g_2}}\frac{\partial g_1}{\partial t} + \frac{\partial^2 f}{\partial y^2}\bigg|_{\substack{x=g_1\\y=g_2}}\frac{\partial g_2}{\partial t}\right).$$

For this example the second partial derivative has six terms in total!

✍️  **Mastery Check 2.20:**

Consider the function   $z = f(x,y) = x \cos y + y$,

where   $x = u(s,t) = st$,   $y = v(s,t) = s^2 + t^2$.   Determine $\dfrac{\partial^2 z}{\partial s \partial t}$.

✍️

**The Leibniz integral rule**

A particularly important application of the chain rule is to differentiating an integral such as

$$\frac{\mathrm{d}}{\mathrm{d}t} \int_{a(t)}^{b(t)} h(x,t)\,\mathrm{d}x,$$

with respect to a parameter. Suppose $h$, $a$ and $b$ are $C^1$ functions and the indicated integral, as well as the integral of $\partial h/\partial t$, exists. The integral itself produces a function $z = F(t) = (f \circ g)(t)$, which depends on $t$ through three channels. Let's call these $u$, $v$ and $t$, where $u$ and $v$ take the place of the upper and lower limits of the integral, and where

$$f : \mathbb{R}^3 \longrightarrow \mathbb{R},$$
$$(u, v, t) \longmapsto z = \int_u^v h(x,t)\,\mathrm{d}x,$$
$$\text{and } \boldsymbol{g} = \big(u = a(t), v = b(t)\big).$$

The branch model relevant to this is shown in Figure 2.36. It is a somewhat simplified version of Case 4 on Page 93.

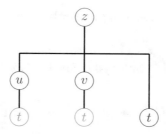

**Figure 2.36**  Ball-and-stick for $f\big(g_1(t), g_2(t), t\big)$.

Referring back to that Case 4, the derivative of $F$ with respect to $t$ is thus

$$\frac{\mathrm{d}F}{\mathrm{d}t} = \left(\frac{\partial f}{\partial u}\right)\bigg|_{u=b(t),v=a(t),t} \cdot \left(\frac{\mathrm{d}b}{\mathrm{d}t}\right)$$
$$+ \left(\frac{\partial f}{\partial v}\right)\bigg|_{u=b(t),v=a(t),t} \cdot \left(\frac{\mathrm{d}a}{\mathrm{d}t}\right) + \left(\frac{\partial f}{\partial t}\right)\bigg|_{u=b(t),v=a(t),t}.$$

The partial derivatives of $f$ with respect to $u$ and $v$ are straightforward using the fundamental theorem of integral calculus:

$$g(x) = \pm\frac{dG}{dx} \iff G(x) = \pm\int_c^x g(x')\mathrm{d}x'$$

where $c$ is some constant. Thus,

$$\left(\frac{\partial f}{\partial u}\right)\bigg|_{(b(t),a(t),t)} = h(b(t),t) \quad \text{and} \quad \left(\frac{\partial f}{\partial v}\right)\bigg|_{(b(t),a(t),t)} = -h(a(t),t).$$

For the partial derivative of $f$ with respect to $t$ we use the definition:

$$\frac{f(u,v,t+\Delta t) - f(u,v,t)}{\Delta t} = \int_u^v \left[\frac{h(x,t+\Delta t) - h(x,t)}{\Delta t}\right]\mathrm{d}x.$$

where we are permitted to put everything under the one integral sign since $f$ is $C^1$. Now, taking the limit $\Delta t \to 0$ we get

$$\frac{\partial f}{\partial t} = \int_u^v \frac{\partial h}{\partial t}(x,t)\,\mathrm{d}x.$$

All together, replacing $u$ and $v$ everywhere with $a(t)$ and $b(t)$, respectively, we have the very useful Leibniz rule

$$\frac{\mathrm{d}}{\mathrm{d}t}\int_{a(t)}^{b(t)} h(x,t)\mathrm{d}x$$

$$= h(b(t),t)\cdot\left(\frac{\mathrm{d}b}{\mathrm{d}t}\right) - h(a(t),t)\cdot\left(\frac{\mathrm{d}a}{\mathrm{d}t}\right) + \int_{a(t)}^{b(t)} \frac{\partial h}{\partial t}(x,t)\mathrm{d}x$$

■ **Example 2.12:**

We apply this rule to the following integrals. Note the use of different independent variables.

(1) Suppose $F(x) = \int_0^{x^2} \sin(u^2)\,\mathrm{d}u$. Then $F'(x) = 2x\sin(x^4)$.

(2) Suppose $F(u) = \int_{1-\ln u}^1 e^{t^2}\,\mathrm{d}t$. Then $F'(u) = \frac{1}{u}e^{(1-\ln u)^2}$.

(3) Suppose $F(t) = \int_{\sin t}^{\cos t} e^{2xt}\,\mathrm{d}x$. Then

$$F'(t) = -e^{2t\cos t}\sin t - e^{2t\sin t}\cos t + \int_{\sin t}^{\cos t} 2xe^{2xt}\,\mathrm{d}x.$$

■

## 2.H   Implicit functions

Suppose we are given the following task: In each of the cases below express the variable $y$ as a function of the remaining variables:

(a) $8y + 64x^2 = 0$;

(b) $2y^2 + 8y + 16z\sin x = 0$;

(c) $\ln|y| + y^3x + 20x^2 = w$.

I am as certain that you cannot complete task (c) as I am that you *can* complete tasks (a) and (b). Although task (c) is impossible, the equation suggests there is a functional relationship, in principle.

This introduces the notion of an implied or *implicit function*. In task (c) the equation implies that $y$ can be a function $f$ of the variables $x$ and $w$. What we shall do in this section is establish conditions under which such a function is defined, *at least locally*. Along the way we will get, as reward, a linear approximation to this unknown function, in terms of the independent variables near a given point, and an explicit expression, and value, for the derivative (or derivatives) of this function at that point.

As before we explain by considering examples of increasing complexity. In each case we will also discuss an analogous linear problem. Since the argument we follow is based on linearization, we hope that the parallels will facilitate reader understanding. The purist reader may frown on the questionable rigour. However, the possibility of greater appreciation for the end result is worth sacrificing some degree of mathematical sophistication.

Suppose we are given the following three problems:

1) $e^{x+y} + xy = 0 \implies F(x,y) = 0$        — a level curve.

2) $e^{x+y+z} - (x+y+z)^2 = 1 \implies F(x,y,z) = 0$    — a level surface.

3) $\begin{cases} e^{x+y+z} - (x+y+z)^2 - 1 = 0 \\ z\sin(xy) - x\cos(zy) = 0 \end{cases} \implies \begin{cases} F(x,y,z) = 0 \\ G(x,y,z) = 0 \end{cases}$
                                                 — a curve of intersection.

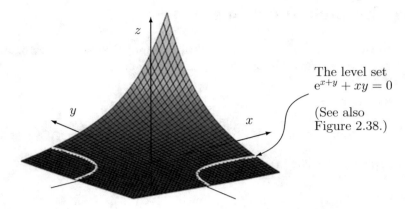

**Figure 2.37** The graph of $z = e^{x+y} + xy$ and the level set $F(x, y) = 0$.

**Consider Problem 1).**   $F(x, y) = e^{x+y} + xy = 0$.

This equation is nonlinear. On top of that, it cannot be manipulated to get $y$ in terms of $x$. All the same, the level curve is shown in Figure 2.37.

But suppose we consider the linear approximation to $F(x, y) = e^{x+y} + xy$ for points $(x, y)$ about a point $(a, b)$ which lies on the level curve $F(x, y) = 0$. The linear approximation is shown in Figure 2.38 and developed on Page 103.

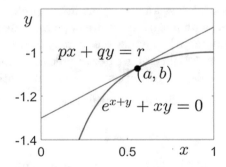

**Figure 2.38** The level curve $e^{x+y} + xy = 0$ and its approximation.

As the analysis parallels the description of a straight line in 2D, it (the general analysis) is preceded by that simple geometric discussion. The student should compare the two mathematical arguments.

Consider the level set describing the general form for the equation of a 2D line (Figure 2.39):

$$ax + by = c, \quad a, b, c \neq 0$$

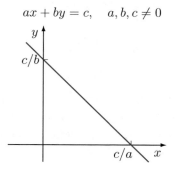

**Figure 2.39** The line $ax + by = c$.

We divide this linear equation by $b$, the coefficient of $y$, and solve for $y$. Naturally, we get the equation of a line in standard form.

$$y = -\frac{a}{b}x + \frac{c}{b} = mx + k,$$

where $-\infty < m = -a/b < \infty$ *as long as* $b \neq 0$. This is an important point for what follows: *The coefficient of the variable we solved for cannot be zero.*

Now consider a more general and potentially nonlinear function, $F$, such as the one suggested here.

If $F \in C^1$, then for all $(x, y)$ *near* $(a, b)$ (see Page 68), we know that a linear approximation to $F$ is obtained from

$$F(x, y) = F(a, b) + \frac{\partial F}{\partial x}\Big|_{(a,b)} .(x - a) + \frac{\partial F}{\partial y}\Big|_{(a,b)} .(y - b) + |\Delta \boldsymbol{x}|\rho(\Delta \boldsymbol{x})$$

where, on the level curve itself, both $F(x, y)$ and $F(a, b)$ are zero:

$$0 = \frac{\partial F}{\partial x}\Big|_{(a,b)} .(x - a) + \frac{\partial F}{\partial y}\Big|_{(a,b)} .(y - b) + |\Delta \boldsymbol{x}|\rho(\Delta \boldsymbol{x}),$$

or, equivalently,

$$\frac{\partial F}{\partial x}\Big|_{(a,b)} .x + \frac{\partial F}{\partial y}\Big|_{(a,b)} .y = \frac{\partial F}{\partial x}\Big|_{(a,b)} .a + \frac{\partial F}{\partial y}\Big|_{(a,b)} .b - |\Delta \boldsymbol{x}|\rho(\Delta \boldsymbol{x}).$$

This is in the general form of the equation of a line $px + qy = r$. Therefore, just as we did in the simple linear case we obtain $y = -(p/q)x + r/q$, provided

$q \neq 0$. This last condition is a key point to remember: the solution for $y$ for points $(x, y)$ near $(a, b)$ is valid *as long as* $q = \partial F / \partial y \neq 0$ at $(a, b)$. And, in that case we have the linear approximation to the curve $F(x, y)$:

$$y = -\frac{\left.\dfrac{\partial F}{\partial x}\right|_{(a,b)}}{\left.\dfrac{\partial F}{\partial y}\right|_{(a,b)}} x + \frac{r}{q}$$

What we are saying is that we have thus obtained an approximate representation for the level curve for points $(x, y)$ near $(a, b)$ that *defines* $y$ locally as a function of $x$. This is given by the straight line in Figure 2.38. (The graph of the actual function is shown in blue.)

Also we have derived the precise value of the derivative of the unknown implicit function $y = f(x)$ at $(a, b)$ even though we cannot write $f$ out explicitly, and in the process we are provided with a condition for the *existence* of the implicit function ($q \neq 0$). The derivative of the implicit function is, in fact,

$$\left.\frac{dy}{dx}\right|_{x=a} = -\frac{\left.\dfrac{\partial F}{\partial x}\right|_{(a,b)}}{\left.\dfrac{\partial F}{\partial y}\right|_{(a,b)}} \iff F \in C^1 \text{ at } (a, b), \text{ and } q = \left.\frac{\partial F}{\partial y}\right|_{(a,b)} \neq 0.$$

### ■ Example 2.13:

Suppose the equation $x^3 y + 2y^3 x = 3$ defines $y$ as a function $f$ of $x$ in the neighbourhood of the point $(1, 1)$. We wish to find the derivative of $f$ at $x = 1$, and a linear approximation to $f$ near the point. Let $F(x, y) = x^3 y + 2y^3 x - 3$.

Note that $F \in C^1 \ \forall \ (x, y) \in \mathbb{R}^2$. Then we have

$$\frac{\partial F}{\partial x} = 3x^2 y + 2y^3, \quad \frac{\partial F}{\partial y} = x^3 + 6xy^2.$$

We note that $\dfrac{\partial F}{\partial y} \neq 0$ at $(1, 1)$. Thus, from our linear approximation we have

$$\left.\frac{dy}{dx}\right|_{(1,1)} = -\frac{3+2}{1+6} = -\frac{5}{7}.$$

The linear approximation is $y = -\dfrac{5}{7}x + c$. To determine $c$, use the fact that the line passes through $(1, 1)$, giving $y = -\dfrac{5}{7}x + \dfrac{12}{7}$.                                                      ■

### ✍ Mastery Check 2.21:

Show that $x^y + \sin y = 1$ defines a function $y = f(x)$ in the neighbourhood of $(1, 0)$, and find $\dfrac{\mathrm{d}y}{\mathrm{d}x}$. Find a linear approximation to $f$, valid near $(1, 0)$.✎

**Consider Problem 2).**   $F(x, y, z) = e^{x+y+z} - (x + y + z)^2 - 1$.

Suppose $\boldsymbol{a} = (a, b, c)$ is a point on the surface. That is, suppose $F(a, b, c) = 0$. We want to know if $F(x, y, z) = 0$ defines a function, $f$, so that the level surface has the form   $z = f(x, y)$   for points $\boldsymbol{x} = (x, y, z)$ on that level surface near $\boldsymbol{a} = (a, b, c)$. In particular, does there exist a tangent plane approximation

$$z = c + \frac{\partial f}{\partial x}\bigg|_{(a,b)} . (x - a) + \frac{\partial f}{\partial y}\bigg|_{(a,b)} . (y - b) \tag{2.3}$$

to the surface at this point? The answer depends on the behaviour of $F(x, y, z)$ near the point, and on the existence of the linear approximation to $F$.

We again lead with an analogy from linear algebra. Consider the equation of the plane shown in Figure 2.40:

$$ax + by + cz = d, \quad a, b, c, d > 0.$$

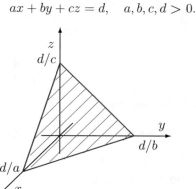

**Figure 2.40**   The plane $ax + by + cz = d$.

We divide this equation by $c$, the coefficient of $z$, and solve for $z$ to get

$$z = \frac{d}{c} - \frac{a}{c}x - \frac{b}{c}y \quad \text{if } c \neq 0.$$

This plane has (partial) slope $-\dfrac{a}{c}$ in the $x$-direction and (partial) slope $-\dfrac{b}{c}$ in $y$-direction; both slopes will be finite as long as $c \neq 0$. This is an important point for our next consideration: *The coefficient of the variable we solve for cannot be zero.* Note the outcome that, subject to this condition, we are able to express one variable, $z$, in terms of the other two variables, $x$ and $y$.

Now let's consider the nonlinear function $F$ of $x, y, z$, supposing that $F$ is differentiable at $\boldsymbol{a}$. Under the latter condition we can obtain the linear approximation to $F$ for points $\boldsymbol{x}$ near $\boldsymbol{a}$ by the methods learned earlier in this chapter:

$$F(x, y, z) = \cancel{F(a, b, c)}^{0} + \left.\frac{\partial F}{\partial x}\right|_{(a,b,c)}.(x - a) + \left.\frac{\partial F}{\partial y}\right|_{(a,b,c)}.(y - b) + \left.\frac{\partial F}{\partial z}\right|_{(a,b,c)}.(z - c)$$
$$+ \, |\boldsymbol{x} - \boldsymbol{a}|\rho(\boldsymbol{x} - \boldsymbol{a}) \text{ (small if } \boldsymbol{x} \text{ near } \boldsymbol{a}) = 0.$$

This approximation can be rearranged to give

$$\left.\frac{\partial F}{\partial z}\right|_{(a,b,c)}.(z - c) = -\left.\frac{\partial F}{\partial x}\right|_{(a,b,c)}.(x - a) - \left.\frac{\partial F}{\partial y}\right|_{(a,b,c)}.(y - b)$$
$$+ \text{ small terms for } (x, y, z) \text{ near } (a, b, c).$$

That is, for $\boldsymbol{x}$ near $\boldsymbol{a}$ we obtain

$$z = c - \frac{\left.\frac{\partial F}{\partial x}\right|_{(a,b,c)}}{\left.\frac{\partial F}{\partial z}\right|_{(a,b,c)}}.(x - a) - \frac{\left.\frac{\partial F}{\partial y}\right|_{(a,b,c)}}{\left.\frac{\partial F}{\partial z}\right|_{(a,b,c)}}.(y - b) \tag{2.4}$$
$$\Longleftrightarrow \left.\frac{\partial F}{\partial z}\right|_{(a,b,c)} \neq 0.$$

This is the tangent plane approximation to $F(x, y, z) = 0$ at $(a, b, c)$.

If $\left.\dfrac{\partial F}{\partial z}\right|_{(a,b,c)} \neq 0$, then the tangent plane is well defined and is identical to the tangent plane to the surface $z = f(x, y)$.

That is, equating corresponding terms in Equations (2.3) and (2.4), we obtain explicit values of the partial derivatives of our implicit function $f$ at $(a, b)$ (with $z = c$).

$$\left.\frac{\partial f}{\partial x}\right|_{(a,b)} = -\frac{\left.\frac{\partial F}{\partial x}\right|_{(a,b,c)}}{\left.\frac{\partial F}{\partial z}\right|_{(a,b,c)}}, \qquad \left.\frac{\partial f}{\partial y}\right|_{(a,b)} = -\frac{\left.\frac{\partial F}{\partial y}\right|_{(a,b,c)}}{\left.\frac{\partial F}{\partial z}\right|_{(a,b,c)}}.$$

So, if $\left.\dfrac{\partial F}{\partial z}\right|_{(a,b,c)} \neq 0$, then both $\left.\dfrac{\partial f}{\partial x}\right|_{(a,b)}$ and $\left.\dfrac{\partial f}{\partial y}\right|_{(a,b)}$ are *well defined*, and $y = f(x, y)$ *exists locally*. We therefore have all that is needed to answer the question posed.

**Consider Problem 3).**
$$\begin{cases} F(x, y, z) = e^{x+y+z} - (x + y + z)^2 - 1 = 0 \\ \quad G(x, y, z) = z \sin(xy) - x \cos(zy) = 0 \end{cases}.$$

The reader who might expect there to be a parallel with a linear algebraic system will not be disappointed to know that we preface the discussion with a review of two pertinent problems in linear algebra.

Suppose we have the situation posed in Figure 2.41. This represents two lines in a plane and a $2 \times 2$ system of equations for unknowns $x$ and $y$.

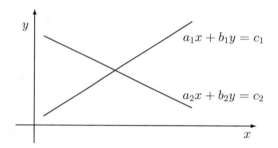

**Figure 2.41**  Two intersecting lines.

When conveniently expressed in matrix form, the system can be readily solved.

$$\begin{pmatrix} a_1 & b_1 \\ a_2 & b_2 \end{pmatrix} \begin{pmatrix} x \\ y \end{pmatrix} = \begin{pmatrix} c_1 \\ c_2 \end{pmatrix} \implies \begin{pmatrix} x \\ y \end{pmatrix} = \frac{1}{a_1 b_2 - a_2 b_1} \begin{pmatrix} b_2 & -b_1 \\ -a_2 & a_1 \end{pmatrix} \begin{pmatrix} c_1 \\ c_2 \end{pmatrix}$$

$$\implies \quad x = \frac{c_1 b_2 - c_2 b_1}{a_1 b_2 - a_2 b_1} = \frac{\begin{vmatrix} c_1 & c_2 \\ b_1 & b_2 \end{vmatrix}}{\begin{vmatrix} a_1 & b_1 \\ a_2 & b_2 \end{vmatrix}}, \quad y = \frac{-c_1 a_2 + c_2 a_1}{a_1 b_2 - a_2 b_1} = \frac{\begin{vmatrix} a_1 & a_2 \\ c_1 & c_2 \end{vmatrix}}{\begin{vmatrix} a_1 & b_1 \\ a_2 & b_2 \end{vmatrix}}$$

We know that this system of equations has a unique solution *if and only if* the determinant of the original matrix is not equal to zero. In that case the solution corresponds to a single point in $\mathbb{R}^2$.

If, on the other hand, the determinant $a_1 b_2 - a_2 b_1 = 0$, then the two equations are linearly dependent and either:

1) the two lines are parallel and no solution exists (the numerator $\neq 0$), or

2) the lines are parallel and superimposed, in which case there are an infinite number of solutions (numerator $= 0$).

As in the earlier linear problems, the condition of a nonzero determinant is the important point to note.

Now consider a second problem from linear algebra, that of two planes.

$$\left.\begin{array}{l} a_1x + b_1y + c_1z = d_1 \\ a_2x + b_2y + c_2z = d_2 \end{array}\right\} \qquad \text{two equations in} \\ \text{three unknowns } (x, y, z).$$

Again, the objective is to determine whether the planes intersect or not, *i.e.* if there exists a solution to the pair of equations. If so, then the solution would correspond to a line of intersection since there are not enough equations to solve for all three variables (unknowns), but we can solve for two of them in terms of the third:

$$\left.\begin{array}{l} a_1x + b_1y = d_1 - c_1z \\ a_2x + b_2y = d_2 - c_2z \end{array}\right\} \implies \begin{pmatrix} a_1 & b_1 \\ a_2 & b_2 \end{pmatrix}\begin{pmatrix} x \\ y \end{pmatrix} = \begin{pmatrix} d_1 - c_1z \\ d_2 - c_2z \end{pmatrix}$$

$$\implies \begin{pmatrix} x \\ y \end{pmatrix} = \frac{1}{a_1b_2 - a_2b_1}\begin{pmatrix} b_2 & -b_1 \\ -a_2 & a_1 \end{pmatrix}\begin{pmatrix} d_1 - c_1z \\ d_2 - c_2z \end{pmatrix}.$$

Again, this system has a unique solution

$$\iff \text{ the determinant} = a_1b_2 - a_2b_1 \neq 0.$$

For example, $\quad x = \dfrac{b_2(d_1 - c_1z) - b_1(d_2 - c_2z)}{a_1b_2 - a_2b_1}, \quad a_1b_2 - a_2b_1 \neq 0,$

$$\implies x = \frac{b_2d_1 - b_1d_2}{a_1b_2 - a_2b_1} - \frac{b_2c_1 - b_1c_2}{a_1b_2 - a_2b_1}.z,$$

$$\left.\begin{array}{l} x = k_1 - m_1z \\ \text{and also } y = k_2 - m_2z \end{array}\right\} \quad \frac{x - k_1}{-m_1} = \frac{y - k_2}{-m_2} = z.$$

As expected this is the equation of a line in 3D (as long as $0 < |m_1|, |m_2| < \infty$).

The two equations for two planes: $\quad \begin{cases} a_1x + b_1y + c_1z = d_1 \\ a_2x + b_2y + c_2z = d_2 \end{cases}$

define a *line of intersection* if the determinant (of the coefficients of the variables we wish to solve for, $x$ and $y$) is nonzero, *i.e.* $a_1b_2 - a_2b_1 \neq 0$. If the determinant is identically zero, then either:

1) the two planes are parallel and there is *no* solution, or

2) the two planes are parallel and superimposed, in which case there are an infinity of solutions.

We could go on to higher dimensional linear manifolds and consider systems of equations of many more variables, but the situations we would encounter would be the same:

• *The determinant of coefficients of the variables to be solved for cannot be zero if we want a unique solution.*

• *If the determinant is zero, then we have either no solution or an infinity of solutions.*

Now that we have reviewed these linear systems we are now ready to tackle the nonlinear problem.

Let $\boldsymbol{a} = (a, b, c)$ be a point on *both* surfaces. That is,

$$F(a, b, c) = G(a, b, c) = 0.$$

Just as in the linear problem on Page 108 these two equations define the set of points $\boldsymbol{x} = (x, y, z)$ which the two surfaces have in common. In other words,

*this set is a curve of intersection.*

Now we ask, when do these equations define a curve that can be expressed

in the form  $\begin{cases} x = f(z), \\ y = g(z), \\ z = z, \end{cases}$  with $z$ as an independent variable?

The answer again depends on the existence of the linear approximations to $F(x, y, z)$ and $G(x, y, z)$ discussed in Section 2.D.

Suppose $F$ and $G$ are differentiable at $\boldsymbol{a} = (a, b, c)$; that is, $F, G \in C^1(\mathbb{R}^3)$. Then for points $\boldsymbol{x}$ near $\boldsymbol{a}$ on the curve of intersection, we find that

$$\begin{cases} F(x, y, z) = \overset{0}{\cancel{F(a, b, c)}} + \dfrac{\partial F}{\partial x}\Big|_a .(x - a) + \dfrac{\partial F}{\partial y}\Big|_a .(y - b) + \dfrac{\partial F}{\partial z}\Big|_a .(z - c) + \cdots = 0 \\[2mm] G(x, y, z) = \overset{0}{\cancel{G(a, b, c)}} + \dfrac{\partial G}{\partial x}\Big|_a .(x - a) + \dfrac{\partial G}{\partial y}\Big|_a .(y - b) + \dfrac{\partial G}{\partial z}\Big|_a .(z - c) + \cdots = 0. \end{cases}$$

Dropping the "$+ \ldots$", these equations can be approximated by

$$\frac{\partial F}{\partial x}\Big|_a .(x-a) + \frac{\partial F}{\partial y}\Big|_a .(y-b) = -\frac{\partial F}{\partial z}\Big|_a .(z-c)$$

$$\frac{\partial G}{\partial x}\Big|_a .(x-a) + \frac{\partial G}{\partial y}\Big|_a .(y-b) = -\frac{\partial G}{\partial z}\Big|_a .(z-c).$$

These linear approximations form a matrix equation for $(x-a)$ and $(y-b)$:

$$\begin{pmatrix} \dfrac{\partial F}{\partial x}\Big|_a & \dfrac{\partial F}{\partial y}\Big|_a \\[2mm] \dfrac{\partial G}{\partial x}\Big|_a & \dfrac{\partial G}{\partial y}\Big|_a \end{pmatrix} \begin{pmatrix} (x-a) \\[2mm] (y-b) \end{pmatrix} = \begin{pmatrix} -\dfrac{\partial F}{\partial z}\Big|_a .(z-c) \\[2mm] -\dfrac{\partial G}{\partial z}\Big|_a .(z-c) \end{pmatrix}$$

This $2 \times 2$ system can be solved for $(x-a)$ and $(y-b)$ *if and only if* the determinant of the coefficient matrix, $\dfrac{\partial F}{\partial x}\Big|_a .\dfrac{\partial G}{\partial y}\Big|_a - \dfrac{\partial F}{\partial y}\Big|_a .\dfrac{\partial G}{\partial x}\Big|_a$, is not identically zero. This is analogous to our second linear algebraic system discussed earlier.

Incidentally, it bears noting that this determinant of derivatives appears in many related but also unrelated contexts (see Chapters 4 and 5 for more details). We take the opportunity here to assign to it a special notation. With it the results to follow are more concisely presented.

---

**Definition 2.9**
*The determinant* $\dfrac{\partial F}{\partial x}\Big|_a .\dfrac{\partial G}{\partial y}\Big|_a - \dfrac{\partial F}{\partial y}\Big|_a .\dfrac{\partial G}{\partial x}\Big|_a \equiv \dfrac{\partial(F,G)}{\partial(x,y)}\Big|_a$ *is called a*
**Jacobian determinant**.

---

Inverting the coefficient matrix and using the Jacobian definition we have

$$(x-a) = -\frac{\dfrac{\partial(F,G)}{\partial(z,y)}\Big|_a}{\dfrac{\partial(F,G)}{\partial(x,y)}\Big|_a}.(z-c), \qquad (y-b) = -\frac{\dfrac{\partial(F,G)}{\partial(x,z)}\Big|_a}{\dfrac{\partial(F,G)}{\partial(x,y)}\Big|_a}.(z-c),$$

for $(x,y,z)$ very near $(a,b,c)$ and provided $\dfrac{\partial(F,G)}{\partial(x,y)}\Big|_a \neq 0$.

Compare these expressions with their linear counterparts in our precursor problem on Page 108.

Therefore, provided $\dfrac{\partial(F,G)}{\partial(x,y)}\Big|_a \neq 0$, the set of equations

$$\begin{cases} x - a = m_1(z - c) \\ y - b = m_2(z - c)\,, \text{ with} \\ z - c = z - c \end{cases} \quad m_1 = -\dfrac{\dfrac{\partial(F,G)}{\partial(z,y)}\Big|_a}{\dfrac{\partial(F,G)}{\partial(x,y)}\Big|_a}, \quad m_2 = -\dfrac{\dfrac{\partial(F,G)}{\partial(x,z)}\Big|_a}{\dfrac{\partial(F,G)}{\partial(x,y)}\Big|_a}$$

is the *tangent line approximation* to the implied curve of intersection

$$x = f(z), \quad y = g(z), \quad z = z, \quad \text{near } (a,b,c). \tag{2.5}$$

We conclude that as long as $m_1$ and $m_2$ are finite the above curve of intersection is well defined. Moreover, since the tangent line to the curve given by Equation (2.5) at the point $(a,b,c)$,

$$\begin{cases} x - a = \dfrac{\mathrm{d}f}{\mathrm{d}z}\Big|_c .(z - c) \\[2mm] y - b = \dfrac{\mathrm{d}g}{\mathrm{d}z}\Big|_c .(z - c) \\[2mm] z - c = z - c, \end{cases}$$

is unique we can also deduce the following results:

$$\dfrac{\mathrm{d}f}{\mathrm{d}z}\Big|_c = -\dfrac{\dfrac{\partial(F,G)}{\partial(z,y)}\Big|_a}{\dfrac{\partial(F,G)}{\partial(x,y)}\Big|_a}, \quad \dfrac{\mathrm{d}g}{\mathrm{d}z}\Big|_c = -\dfrac{\dfrac{\partial(F,G)}{\partial(x,z)}\Big|_a}{\dfrac{\partial(F,G)}{\partial(x,y)}\Big|_a}.$$

Once again, the conditions for these to be valid and for the implicit functions to exist are

(i)  *F and G differentiable at* $(a,b,c)$;

(ii)  *the Jacobian determinant* $J = \dfrac{\partial(F,G)}{\partial(x,y)} \neq 0$ *at* $(a,b,c)$.

The important point to note in this example is that critical condition (ii) involves the matrix of coefficients of the dependent variables, $x$ and $y$. This is consistent with the preceding examples and is a general rule of thumb with establishing the existence of any implicit functions!

## ✍ Mastery Check 2.22:

Suppose $x = h(u, v) = u^2 + v^2$ and $y = k(u, v) = uv$ are to be "solved" for $u$ and $v$ in terms of $x$ and $y$. Find $\dfrac{\partial u}{\partial x}$, $\dfrac{\partial u}{\partial y}$, $\dfrac{\partial v}{\partial x}$, $\dfrac{\partial v}{\partial y}$, where possible. Show

$$\frac{\partial(u, v)}{\partial(x, y)} = \left( \frac{\partial(x, y)}{\partial(u, v)} \right)^{-1}$$ provided the denominator $\neq 0$. Hint: This problem

is similar to Problem 3) above, but this time with four variables. It proves to be worthwhile here to be a little liberal with our notation convention, and refer to $u$ and $v$ as functions as well as variables.

1) Define suitable level sets $\begin{cases} F(x, y, u, v) = 0, \\ G(x, y, u, v) = 0, \end{cases}$ which we wish to solve
   for some functions $u = f(x, y)$, $v = g(x, y)$, the curves of intersection.

2) Set up the linear approximations to the level sets for $(x, y, u, v)$ near $(a, b, c, d)$. Rewrite the approximations as a matrix equation in $\begin{pmatrix} u - c \\ v - d \end{pmatrix}$.

3) Solve the equation. (What appears as a denominator?)

4) Compare your solution to the true tangent lines and so obtain $\dfrac{\partial f}{\partial x}$, $\dfrac{\partial f}{\partial y}$, $\dfrac{\partial g}{\partial x}$, $\dfrac{\partial g}{\partial y}$, in terms of Jacobians.

5) Finally, compare $\dfrac{\partial(f, g)}{\partial(x, y)}$ with $\dfrac{\partial(h, k)}{\partial(u, v)}$.

✍

In this last Mastery Check, we met a *fundamental property* of the Jacobian

(where it exists), namely,     $\dfrac{\partial(u, v)}{\partial(x, y)} = \left( \dfrac{\partial(x, y)}{\partial(u, v)} \right)^{-1}$.

In the single-variable case, it is true that $\dfrac{dy}{dx} = 1 \Big/ \dfrac{dx}{dy}$ if $\dfrac{dx}{dy} \neq 0$.

However in the context of multivariable functions and partial derivatives it is the *Jacobian* which takes the place of the ordinary derivative.

In the notation used in the Mastery Check,

$$\frac{\partial u}{\partial x} \left( = \frac{\partial f}{\partial x} \right) \neq 1 \Big/ \frac{\partial x}{\partial u} \left( = 1 \Big/ \frac{\partial h}{\partial u} \right).$$

✍  **Mastery Check 2.23:**

Let $\begin{cases} F(x, y, u, v) = xyuv - 1 = 0 \\ G(x, y, u, v) = x + y + u + v = 0, \end{cases}$

and consider points $P_0 = (1, 1, -1, -1)$, $P_1 = (1, -1, 1, -1)$.

Find $\left(\dfrac{\partial y}{\partial x}\right)_u$ at $P_0$ and $P_1$.

Hint: Which are the independent variables?

✍

✍  **Mastery Check 2.24:**

Show that the system of equations $\begin{cases} xy^2 + xzu + yv^2 = 3 \\ x^3yz + 2xv - u^2v^2 = 2 \end{cases}$ can be solved

for $u, v$ as functions of $x, y, z$ near the point $P_0(1, 1, 1, 1, 1)$.

Find the value of $\dfrac{\partial v}{\partial y}$ for the solution at $(x, y, z) = (1, 1, 1)$.

✍

# 2.I  Taylor's formula and Taylor series

On the one hand, this next topic is but a natural extension of our earlier discussion on tangent plane approximations (Section 2.D). On the other hand, the subject of Taylor series and Taylor approximations is so incredibly useful in analysis and incredibly practical in computational applications that it is worth giving some consideration. The next chapter will highlight some examples of such applications. For the present purpose we consider this topic as a means of getting good approximations to functions, whether explicit or implicit. A convenient place to start is with the single-variable case.

Recall the properties of Taylor and Maclaurin polynomials for any function $F : \mathbb{R} \longrightarrow \mathbb{R}$: Let $F$ and $F^{(k)}$, $k = 1, 2, \ldots, n$ be continuous on an open interval $I$ including the point $t_0$, and let $F^{(n+1)}(t)$ exist for all $t \in I$. The best polynomial approximation of order $n$ to $f$ near $t_0 \in I$ is the first contribution in the next equation:

$$\begin{aligned} F(t) &= P_n(t) + E_n(t; t_0) \\ &= F(t_0) + F'(t_0)(t - t_0) + \frac{F''(t_0)}{2}(t - t_0)^2 + \cdots \\ &\quad + \frac{F^{(n)}(t_0)}{n!}(t - t_0)^n + E_n(t; t_0). \end{aligned}$$

$P_n(t)$ is referred to as the Taylor polynomial approximation to $F$ of order $n$, while

$$E_n(t; t_0) = F(t) - P_n(t)$$
$$= \frac{F^{(n+1)}(a)}{(n+1)!}(t - t_0)^{n+1}, \qquad a \in \big(\min(t_0, t), \max(t_0, t)\big)$$

is the error term (the difference between the true value and its approximation).

Here are a few special cases for $n < \infty$:

(i) $P_1(t) = F(t_0) + F'(t_0)(t - t_0)$          — linear approximation

(ii) $P_2(t) = P_1(t) + \dfrac{F''(t_0)}{2}(t - t_0)^2$     — quadratic approximation

(iii) $P_3(t) = P_2(t) + \dfrac{F'''(t_0)}{3!}(t - t_0)^3$     — cubic approximation

(iv) $P_4(t) = P_3(t) + \dfrac{F^{(4)}(t_0)}{4!}(t - t_0)^4$     — quartic approximation

**Specific cases to note**

* The existence of a linear approximation means there is a tangent line to $F(t)$ at $t = t_0$.

* A quadratic approximation is useful for critical-point analysis when $F'(t_0) = 0$, meaning that

$$F(t) - F(t_0) \approx \frac{F''(t_0)}{2}(t - t_0)^2 \quad \begin{array}{ll} > 0, & \text{— a minimum point} \\ < 0, & \text{— a maximum point.} \end{array}$$

* A cubic approximation means that there is a cubic curve osculating $F(t)$ at $x = t_0$.                             (What is that?)

* A quartic approximation may be useful in the uncommon cases when $F'(t_0) = F''(t_0) = F'''(t) = 0$, meaning that

$$F(t) - F(t_0) \approx \frac{F^{(4)}(t_0)}{4!}(t - t_0)^4 \quad \begin{array}{ll} > 0, & \text{— a minimum point} \\ < 0, & \text{— a maximum point.} \end{array}$$

Some functions have derivatives of all orders. Therefore, we can consider extending $n$ without limit, that is, $n \to \infty$. The above polynomial can then be developed to an infinite power series, provided it converges (absolutely) in some interval.

Functions that can be differentiated an indefinite number of times and whose interval of convergence is the whole real line, for example, $\sin t$, $\cos t$, $e^t$, are called *analytic*.

The Taylor series representation of a function $F(t)$ is defined as follows:

---

**Definition 2.10**

*If there exists a $t_0 \in \mathbb{R}$ and an $R > 0$ such that $F(t) = \sum_{k=0}^{\infty} a_k (t - t_0)^k$*

*converges for $|t - t_0| < R$, then this is the Taylor series of $F$,*

*and $a_k = \dfrac{F^{(k)}(t_0)}{k!}$ $\forall k$.*

---

What is especially important for us is the particular choice $t_0 = 0$. For this choice we have the well-known *Maclaurin polynomial*.

To be precise, if a single-variable function $F : \mathbb{R} \longrightarrow \mathbb{R}$ has continuous derivatives of all orders less than or equal to $n + 1$, in an open interval $I$ centred at $t = 0$, then $F$ can be approximated by the Maclaurin polynomial,

$$F(t) = F(0) + F'(0)t + \frac{1}{2!}F''(0)t^2 + \cdots + \frac{1}{n!}F^{(n)}(0)t^n + R_n(\theta, t) \quad (2.6)$$

for all $t$ in the interval $I$, and where

$$R_n(\theta, t) = \frac{1}{(n+1)!}F^{(n+1)}(\theta)t^{n+1}$$

is the measure of error in the approximation, with $0 < |\theta| < t$.

As with the Taylor series, if $F$ has derivatives of all orders, that is, for $n \to \infty$, at $t = 0$, then we can define the Maclaurin series representation of $F$:

$$F(t) = \sum_{k=0}^{\infty} a_k t^k \text{ for } t \in I, \text{ with } 0 \in I, \text{ and } a_k = \frac{F^{(k)}(0)}{k!} \ \forall k.$$

For example, it is easy to verify by repeated differentiation and substitution that the Maclaurin series for the sine function is

$$\sin t = \sum_{k=0}^{\infty} \frac{(-1)^{k+1}}{(2k+1)!} t^{2k+1}.$$

From this single-variable case we can derive corresponding versions of Taylor and Maclaurin polynomials for functions of several variables.

Let's consider a special function $F$ and look at its value at $t = 1$:

Let $f : \mathbb{R}^2 \longrightarrow \mathbb{R}$ be defined and have continuous partial derivatives of orders $0, 1, \ldots, n+1$, at the point $(x_0, y_0)$ in the domain of $f$.
For fixed $(x, y) \in S_r(x_0, y_0)$ and therefore fixed $h = x - x_0$ and $k = y - y_0$, consider $f(x, y)$ at $x = x_0 + th, y = y_0 + tk$. We can therefore define a single-valued function of $t$

$$F(t) = f(x_0 + th, y_0 + tk)$$

whose value at $t = 0$ is $F(0) = f(x_0, y_0)$, and whose value at $t = 1$ is $F(1) = f(x_0 + h, y_0 + k)$.

We now develop the Maclaurin polynomial for $F$ *via the chain rule*, which leads us to the Taylor polynomial for $f$.

The terms in the Maclaurin polynomial are

$$F(0) = f(x_0, y_0), \qquad F'(0)t = \left( \frac{\partial f}{\partial x} \bigg|_{(x_0, y_0)} h + \frac{\partial f}{\partial y} \bigg|_{(x_0, y_0)} k \right) t,$$

$$\frac{1}{2} F''(0)t^2 = \frac{1}{2} \left( \frac{\partial^2 f}{\partial x^2} \bigg|_{(x_0, y_0)} h^2 + 2 \frac{\partial^2 f}{\partial x \partial y} \bigg|_{(x_0, y_0)} hk + \frac{\partial^2 f}{\partial y^2} \bigg|_{(x_0, y_0)} k^2 \right) t^2 \ldots .$$

Inserting these in (2.6) and letting $t = 1$ give us the Taylor polynomial approximation to $f$:

$$\begin{aligned}
f(x_0 + h, y_0 + k) = {} & f(x_0, y_0) + \frac{\partial f}{\partial x} \bigg|_0 h + \frac{\partial f}{\partial y} \bigg|_0 k \\
& + \frac{1}{2} \left( \frac{\partial^2 f}{\partial x^2} \bigg|_0 h^2 + 2 \frac{\partial^2 f}{\partial x \partial y} \bigg|_0 hk + \frac{\partial^2 f}{\partial y^2} \bigg|_0 k^2 \right) + \cdots \\
& + \sum_{j=0}^{n} \frac{1}{j!(n-j)!} \frac{\partial^n f}{\partial x^j \partial y^{n-j}} \bigg|_0 h^j k^{n-j} + R_n(\theta, h, k).
\end{aligned} \qquad (2.7)$$

Alternative derivation:
Suppose we can approximate $f(x, y)$ in a neighbourhood $S_r(x_0, y_0)$ of $(x_0, y_0)$

by a general polynomial:

$$f(x, y) = a_{00} + a_{10}(x - x_0) + a_{01}(y - y_0)$$
$$+ a_{20}(x - x_0)^2 + a_{11}(x - x_0)(y - y_0) + a_{02}(y - y_0)^2 + \cdots$$
$$+ a_{n0}(x - x_0)^n + \cdots + a_{0n}(y - y_0)^n + E_n(\boldsymbol{x}_0).$$

(Approximation error will depend on $\boldsymbol{x}_0$ and $n$.)

If $f$ is differentiable to order $n$, then taking partial derivatives of both sides and evaluating these results at $(x_0, y_0)$ we identify

$$\left.\frac{\partial f}{\partial x}\right|_0 = a_{10} \qquad \left.\frac{\partial f}{\partial y}\right|_0 = a_{01}$$

$$\left.\frac{\partial^2 f}{\partial x^2}\right|_0 = 2a_{20} \qquad \left.\frac{\partial^2 f}{\partial y^2}\right|_0 = 2a_{02} \qquad \left.\frac{\partial^2 f}{\partial x \partial y}\right|_0 = a_{11} = \left.\frac{\partial^2 f}{\partial y \partial x}\right|_0$$

and generally $\left.\dfrac{\partial^{k+\ell} f}{\partial x^k \partial y^\ell}\right|_0 = k!\ell! a_{k\ell}.$

Substitution will give (2.7) again.

We end this section with two Mastery Check exercises involving Taylor polynomials, postponing until the next chapter a demonstration of the usefulness of Taylor approximations. However, it is appropriate first to make a few important comments.

(1) With the trivial step of setting $x_0 = y_0 = 0$ we get the 2D version of a Maclaurin polynomial approximation of order $n$.

(2) As in the single-variable case, if our multivariable function has partial derivatives of all orders, then the Taylor polynomial approximations of our $f(x, y)$ can be developed into a series representation — provided it converges, of course.

(3) In Section 2.H we dealt with implicit functions and showed how one can calculate first derivatives of such functions even though the functions themselves could not be expressed explicitly. Now, with assistance from the Taylor and Maclaurin expansions, one can construct explicit polynomial or even full series representations of implicit functions by successively differentiating these first-order derivatives (with the help of the chain rule — Section 2.G). Our second Mastery Check exercise explores this possibility.

✍  **Mastery Check 2.25:**

Determine Taylor's polynomial of order 2 to the function
$$f(x, y) = \ln(2x^2 + y^2) - 2y$$
about the point $(0, 1)$, and evaluate an approximation to $f(0.1, 1.2)$.

✍

✍  **Mastery Check 2.26:**

Determine the Taylor polynomial of order 2 about the point $(0, 0)$ to the implicit function $z = f(x, y)$ defined by the equation
$$e^{x+y+z} - (x + y + z)^2 - 1 = 0.$$

Hint: see Section 2.H Problem 2.

✍

# 2.J  Supplementary problems

### Section 2.A

1. Consider the function $f : \mathbb{R} \longrightarrow \mathbb{R}$, $x \mapsto f(x)$, where
   (a) $f(x) = x^{2/3}$;        (b) $f(x) = x^{4/3}$;        (c) $f(x) = x^{1/2}$.

   Decide in each case whether the derivative of $f$ exists at $x = 0$ as follows:

   (i) Check whether $f$ exists at $x=0$ and at $x = 0+h$ (for $h$ arbitrarily small).

   (ii) Check right and left limits to see whether $f$ is continuous at $x = 0$.

   (iii) Check right and left limits to see whether $f$ has a derivative at $x = 0$.

2. Consider the function $f : \mathbb{R} \longrightarrow \mathbb{R}$, $x \mapsto (x-1)^\alpha$, $\alpha \in \mathbb{R}$. Decide from first principles for what values of $\alpha$ the derivative of $f$ exists at $x = 1$.

### Section 2.B

3. Are the following functions continuous at $(0,0)$?

   (a) $f(x, y) = \dfrac{xy}{x^2 + y^2}$, $f(0,0) = 0$.

   (b) $f(x, y) = \dfrac{xy}{\sqrt{x^2 + y^2}}$, $f(0,0) = 0$.

4. Establish whether the limits of the following functions exist at the given points, and if so determine their values.

   (a) $f(x, y) = \dfrac{\tan(xy)}{x^2 + y^2}$, at $(x, y) = (0,0)$.

   (b) $f(x, y) = \dfrac{x^2 - x}{y^2 - y}$, at $(x, y) = (0,0)$.

   (c) $f(x, y) = \dfrac{x^4 + y \sin(x^3)}{x^4 + y^4 + x^2 y^2}$, at $(x, y) = (0,0)$.

   (d) $f(x, y) = \dfrac{x^2 \sin y}{x^2 + y^2}$, at $(x, y) = (0,0)$.

5. Use the limit definition to show that

(a) $\lim_{(x,y)\to(1,1)} \left( x + \dfrac{1}{y} \right) = 2.$

(b) $\lim_{(x,y)\to(1,2)} \left( x^2 + 2y \right) = 5.$

(c) $\lim_{(x,y)\to(0,0)} \left( \dfrac{\cos(2xy)}{1 + x^2 + y^2} \right) = 1.$

Repeat the limit calculation using the limit laws, where these are applicable.

## Section 2.C

6. Find all first partial derivatives of the following functions:

(a) $f(x,y) = \arctan(y/x).$

(b) $f(x,y) = \arctan(x^2 + xy + y^2).$

(c) $f(x,y) = \left( x^2 y^3 + 1 \right)^3.$

(d) $f(x,y) = \exp\left( x^2 \sin^2 y + 2xy \sin x \sin y + y^2 \right).$

(e) $f(x,y,z) = x^2 \sqrt{y^2 + xz}.$

(f) $f(x_1, x_2, x_3) = \ln |x_1 x_2 + x_2 x_3 + x_1 x_3|.$

## Section 2.D

7. Consider the function $f : \mathbb{R}^2 \longrightarrow \mathbb{R}$, $f(x,y) = x\arctan(y/x)$,
$f(0,y) = 0.$
Discuss (a) the continuity of $f$, and (b) the continuity and existence of $f_x$ and $f_y$, at points on the $y$-axis, without first drawing the graph.
Discuss (c) the existence of the second partial derivatives at these points, and determine them if they do.

## Section 2.E

8. Consider the function $f(x,y,z) = xy^3 + yz^2$. What is the directional derivative at the point $(1, 2, -1)$ in the direction $\boldsymbol{u} = 2\boldsymbol{e}_1 + \boldsymbol{e}_2 + 2\boldsymbol{e}_3$?

9. Consider the surface $xy^3 z^2 = 8$. What is the unit normal to this surface at the point $(1, 2, -1)$?

10. Determine the points on the flattened ellipsoid
$$x^2 + 13y^2 + 7z^2 - 6\sqrt{3}zy = 16$$
where the tangent plane is parallel to one of the coordinate planes.

11. Determine all points on the surface
$$x^2 + 2y^2 + 3z^2 + 2xy + 2yz = 1$$
where the tangent plane is parallel to the plane $x + y + z = 0$.

## Section 2.F

12. Find the first and second partial derivatives of the following functions:

    (a) $f(x, y) = \left( \sin xy \right)^2$.

    (b) $f(x, y) = \ln \left( x^2 + 2y \right)$.

    (c) $f(x, y) = \sqrt{1 + x^2 + y^3}$.

13. Suppose $f : \mathbb{R}^2 \to \mathbb{R}$ is a $C^3$ harmonic function of variables $(x, y)$. That is, suppose $f$ satisfies the 2D Laplace equation,
$$\Delta f = \frac{\partial^2 f}{\partial x^2} + \frac{\partial^2 f}{\partial y^2} = 0$$
(see Section 3.E for more information on this topic). Show that the function $g = x\dfrac{\partial f}{\partial x} + y\dfrac{\partial f}{\partial y}$ is also a solution.

## Section 2.G

14. Consider the functions $f : x \mapsto y = f(x)$, and $g : t \mapsto x = g(t)$. In each example that follows, find the domain $D_F$ of the composite function $F : t \mapsto y = (f \circ g)(t)$, and the derivative $\dfrac{\mathrm{d}F}{\mathrm{d}t}$.

    (a) $f(x) = \dfrac{x^2}{1 + x^2}$, $g(t) = \sinh t$.
    (Do this example in two ways: by finding the expression for $F(t)$ explicitly in terms of $t$, then differentiating; and by using the chain rule.)

    (b) $f(x) = \arcsin x^2$, $g(t) = 3\mathrm{e}^{-t}$.

15. Consider the functions $f : x \mapsto y = f(x)$, and $g : (s,t) \mapsto x = g(s,t)$. In each example that follows, find the domain $D_F$ of the composite function $F : (s,t) \mapsto y = (f \circ g)(s,t)$, and the derivatives $\dfrac{\partial F}{\partial s}, \dfrac{\partial F}{\partial t}$.

   (a) $f(x) = \ln x, \quad g(s,t) = s(1 - t^2)$.

   (b) $f(x) = \arccos(x), \quad g(s,t) = \sqrt{s^2 - t^2}$.

16. Consider the function $z = f(x,y) = e^{x^2 y} + xy$, and the composite function $z = F(t) = f(\cos t, \sin t)$. Decide whether $F(t)$ makes sense, and if so, find its domain and compute $\dfrac{dF}{dt}$.

17. Consider the function $z = f(x,y) = \arcsin xy$, where $x = s - 2t$ and $y = \dfrac{s}{t^2}$.

   Check that the composite function $z = F(s,t)$ makes sense, and if so, find its domain and compute $\dfrac{\partial F}{\partial s}$ and $\dfrac{\partial F}{\partial t}$.

18. By introducing new variables $u = x^2 - y$ and $v = x + y^2$ transform the differential equation

$$(1 - 2y)\frac{\partial f}{\partial x} + (1 + 2x)\frac{\partial f}{\partial y} = 0.$$

19. Suppose $f : \mathbb{R}^2 \to \mathbb{R}$ is a $C^2$ function of variables $(x,y)$. By introducing the change of variables $x = 2s + 3t$ and $y = 4s - 4t$ we define the $C^2$ function $F(s,t)$ from $f(x,y)$. Show that

$$\frac{\partial^2 F}{\partial t^2} = 9\frac{\partial^2 f}{\partial x^2} - 24\frac{\partial^2 f}{\partial y \partial x} + 16\frac{\partial^2 f}{\partial y^2}.$$

20. Suppose $f : \mathbb{R}^2 \to \mathbb{R}$ is a $C^2$ harmonic function of variables $(x,y)$. By introducing 2D polar coordinates $x = r\cos\theta$ and $y = r\sin\theta$ show that Laplace's equation becomes

$$\Delta F = \frac{\partial^2 F}{\partial r^2} + \frac{1}{r}\frac{\partial F}{\partial r} + \frac{1}{r^2}\frac{\partial^2 F}{\partial \theta^2} = 0$$

   where $F(r,\theta) = f(r\cos\theta, r\sin\theta)$.

21. Suppose $f : \mathbb{R}^2 \to \mathbb{R}$ is a $C^2$ function of variables $(x,y)$, by introducing the new variables $s = x^2 + y$ and $t = 2x$ transform the expression

$$\frac{\partial^2 f}{\partial x^2} + 2\frac{\partial^2 f}{\partial y \partial x} + \frac{\partial^2 f}{\partial y^2}$$

into a form involving a function $F(s,t)$.

## Section 2.H

22. Suppose the equation $x^3y + 2y^3x = 3$ defines $y$ as a function $f$ of $x$ in the neighbourhood of the point $(1,1)$. Find the derivative of $f$ at $x = 1$.

23. Suppose the equation $z^3 + z(y^2 + 1) + x^3 - 3x + y^2 - 8 = 0$ defines $z$ as a function $f$ of $x, y$ in the neighbourhood of the point $(2, -1, -1)$. Find the derivatives of $\dfrac{\partial f}{\partial x}$ and $\dfrac{\partial f}{\partial y}$ at this point.

## Section 2.I

24. Find the Taylor polynomial approximations of order 1, 2, ..., to the function $F(t) = \cos(t^{3/2})$ about the point $t = t_0 = 0$. Use a suitable low-order approximation to determine whether $t = 0$ is a maximum or minimum point.

25. If a function may be approximated by a convergent Taylor series whose terms are alternating in sign, then the absolute value of the error term is bounded by the absolute value of the first omitted term in a finite polynomial approximation. Use this idea to decide how many terms are needed to find $\sin 3°$ to ten decimal places.

26. Write down the Taylor series for $\cos t$ about $t_0 = \pi/3$, and use it to compute $\cos 55°$ to seven decimal places. (First establish how many terms are needed.)

27. Write down all the terms up to order $n = 3$ in the Taylor series for
$$f(x,y) = \sin(x+y) + x^2 \text{ about } (x_0, y_0) = (0, \pi/6).$$

28. The function $z = f(x,y)$ satisfies $f(1,1) = 0$ and $e^{xz} + x^2y^3 + z = 2$. Establish that $z = f(x,y)$ indeed exists in a region near $(1.1)$ and determine its Taylor series approximation up to order 2 valid near that point.

29. Determine the Taylor polynomial approximation up and including second-order terms of $F(x,y) = \displaystyle\int_0^y \ln(1 + x + t)\,dt$ about the point $(0,0)$. Note that we assume $x + y > -1$.

# Chapter 3

# Applications of the differential calculus

In this chapter, the theory of multivariable functions and their partial derivatives as covered in the preceding chapter is applied to problems arising in four contexts: finding function maxima and minima, error analysis, least-square approximations, and partial differential equations. Although applications arise in a wide variety of forms, these are among the more common examples.

## 3.A  Extreme values of $f : \mathbb{R}^n \longrightarrow \mathbb{R}$

Occupying a central position in the vastness of the space of applications of differential calculus is the subject of *optimization*. At its most basic, the term refers to the task of finding those points in a function's domain that gives rise to maxima or minima of that (scalar) function, and of determining the corresponding values of that function.

Of special interest in the study are the so-called *extreme points* of $f(\boldsymbol{x})$, a subset of which are the so-called *critical points*. These are points where the function can exhibit either a local maximum or minimum, and even a global maximum or minimum.

To set the stage we require some basic infrastructure. We start with a few essential definitions.

© Springer Nature Switzerland AG 2020
S. J. Miklavcic, *An Illustrative Guide to Multivariable and Vector Calculus*,
https://doi.org/10.1007/978-3-030-33459-8_3

**Definition 3.1**
*Consider a continuous $f : \mathbb{R}^n \longrightarrow \mathbb{R}$.*
*A point $\boldsymbol{a} \in D_f$ is called a* **local**
        *(i)* **minimum point** *if $f(\boldsymbol{x}) \geq f(\boldsymbol{a})$, $\forall \boldsymbol{x} \in S_r(\boldsymbol{a})$,*
        *(ii)* **maximum point** *if $f(\boldsymbol{x}) \leq f(\boldsymbol{a})$, $\forall \boldsymbol{x} \in S_r(\boldsymbol{a})$.*

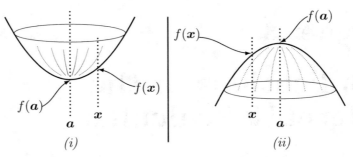

**Figure 3.1** *A function minimum and maximum.*

We have here invoked the open sphere $S_r(\boldsymbol{a})$ to represent the set of points $\boldsymbol{x}$ different from but near $\boldsymbol{a}$ (the radius $r > 0$ is presumed small). We could equally well have referred to points $\boldsymbol{x}$ in a larger "neighbourhood" of $\boldsymbol{a}$. However, that proves to be unnecessary and less convenient, it is enough to consider a small open sphere as we are defining *local* properties.

Points of local minimum (Figure 3.1(i)) and local maximum (Figure 3.1(ii)) are examples of *critical points*.

**Definition 3.2**
*A* **critical point** *is an interior point $\boldsymbol{a} \in D_f$ at which $\boldsymbol{\nabla} f\big|_{\boldsymbol{a}} = \boldsymbol{0}$ for $f \in C^1$.*

While a local maximum point or a local minimum point must mean that $\boldsymbol{\nabla} f = \boldsymbol{0}$ *at that point* the converse is not necessarily true, a critical point where $\boldsymbol{\nabla} f = \boldsymbol{0}$ need *not* be either a point of maximum or minimum; there is a third alternative.

> **Definition 3.3**
> *A critical point which is neither a maximum nor a minimum is called a*
> **saddle point**.

Referring to Definition 3.1, in the case of a local minimum (left-hand figure) or a local maximum (right-hand figure) at $\boldsymbol{x} = \boldsymbol{a}$, the tangent plane is horizontal, which is a consequence of $\boldsymbol{\nabla} f = \boldsymbol{0}$ at $\boldsymbol{a}$.

For a saddle point the tangent plane is still horizontal but neither of the figures in Definition 3.1 applies. Instead, around a saddle point a part of the function's graph is below the tangent plane and a part is above.

The following two simple examples convey the general idea of the above definitions.

■  **Example 3.1:**
Consider the function $z = f(x, y) = x^2 + y^2 - 2x$. We have

$$\boldsymbol{\nabla} f \; = \; \begin{pmatrix} 2x - 2 \\ 2y \end{pmatrix} \; = \; \begin{pmatrix} 0 \\ 0 \end{pmatrix} \text{ at } (x, y) = (1, 0).$$

Now we examine $f$ in the neighbourhood of this critical point $(1, 0)$. (Note that there is just one critical point in this example.) Let's consider the neighbouring point $(1 + h, 0 + k)$ in the domain of $f$. We have

$$f(1 + h, 0 + k) \; = \; (1 + h)^2 + k^2 - 2(1 + h) \; = \; h^2 + k^2 - 1,$$

while $f(1, 0) = 1 + 0 - 2 = -1$. We see that $f(1 + h, 0 + k) > f(1, 0)$ for all $h, k \neq 0$, since

$$f(1 + h, 0 + k) - f(1, 0) = h^2 + k^2 > 0.$$

As this is true for all $(h, k)$, that is, all $(x, y)$ in the neighbourhood of $(1, 0)$, the point $(1, 0)$ is a minimum point.  ■

■  **Example 3.2:**
Consider the function $z = f(x, y) = 1 + x^2 - y^2$. (See Example 1.10.) We have

$$\boldsymbol{\nabla} f \; = \; \begin{pmatrix} 2x \\ -2y \end{pmatrix} \; = \; \begin{pmatrix} 0 \\ 0 \end{pmatrix} \text{ at } (x, y) = (0, 0).$$

We examine $f$ in the neighbourhood of this critical point. At the neighbouring point $(0 + h, 0 + k)$ we have

$$f(0 + h, 0 + k) \; = \; 1 + h^2 - k^2,$$

which is $> 0$ along the line $y = k = 0$, but is $< 0$ along the line $x = h = 0$.
The critical point $(0, 0)$ is neither a local maximum nor a local minimum. It
is a *saddle point*.                                                          ∎

The reader should now try their hand at a similar style problem.

### ✍ Mastery Check 3.1:

Consider the function $z = f(x, y) = x^2 - y^2 - 2x$. Find the point $(a, b)$ at
which $\nabla f = \mathbf{0}$, and then find an expression for $f(a + h, b + k)$ for small
$h, k \neq 0$. Use Definition 3.1 to decide whether the point $(a, b)$ is a maximum or a minimum (or neither).                                          ✍

**Extreme values in two dimensions—general procedure**

The functions in Example 3.1 and the last Mastery Check were nice ones to
work with. We could use simple algebra to determine if a point of interest
was a maximum point or a minimum point or neither. The question naturally
arises, what do we do with functions which are more complicated?

The answer relies on the fact that since we are interested only in *local* extreme
points we can make use of *local approximations* to functions.

In fact, all we usually ever need is Taylor's polynomial of second order which,
as the next theorem states, is enough to represent a function locally.

---

**Theorem 3.1**

*Let $f : D_f \subset \mathbb{R}^2 \longrightarrow \mathbb{R}$ be a function with continuous derivatives of order
0, 1, 2, and 3 (that is, $f$ is a class $C^3$ function) in some neighbourhood
of a point $\mathbf{a} \in D_f$. Then,*

$$f(a + h, b + k) = f(a, b) + \frac{\partial f}{\partial x}(a, b).h + \frac{\partial f}{\partial y}(a, b).k$$

$$+ \frac{1}{2}\left(\frac{\partial^2 f}{\partial x^2}(a, b).h^2 + 2\frac{\partial^2 f}{\partial x \partial y}(a, b).hk + \frac{\partial^2 f}{\partial y^2}(a, b).k^2\right) + (h^2 + k^2)^{3/2}B(h, k),$$

*where $B$ is some bounded function in the neighbourhood of $(0, 0)$.*

---

Using Taylor polynomials of order 2 (see Equation 2.7) results in a considerable simplification. The difficulties of critical point problems involving more
complex functions are reduced to the level featured in Examples 3.1 and 3.2,

since the function approximations are algebraic.

The Taylor polynomial approximation of order 2 can be written more succinctly in a vector-matrix product form

$$f(\boldsymbol{a} + \boldsymbol{h}) = f(\boldsymbol{a}) + \text{grad } f(\boldsymbol{a}) \cdot \boldsymbol{h} + \frac{1}{2}\boldsymbol{h}^T \cdot H_{\boldsymbol{a}}f \cdot \boldsymbol{h} + \text{small terms}, \qquad (3.1)$$

where

$$H_{\boldsymbol{a}}f = \begin{pmatrix} \dfrac{\partial^2 f}{\partial x^2}(a,b) & \dfrac{\partial^2 f}{\partial x \partial y}(a,b) \\ \dfrac{\partial^2 f}{\partial y \partial x}(a,b) & \dfrac{\partial^2 f}{\partial y^2}(a,b) \end{pmatrix}$$

is a symmetric matrix for $C^2$ functions called *the Hessian matrix*, and

$$\boldsymbol{h} = \begin{pmatrix} h \\ k \end{pmatrix}, \quad \boldsymbol{h}^T = (h, k), \quad \text{grad } f(\boldsymbol{a}) = \left( \frac{\partial f}{\partial x}(a,b), \frac{\partial f}{\partial y}(a,b) \right).$$

The vector-matrix expression, Equation 3.1, is a very convenient form to use as it is straightforward to generalize to functions of $n$ variables. For $n \geq 3$ only the sizes of the vectors and the Hessian matrix increases, while the form stays the same. Try it?

We shall now see how Taylor's second-order polynomial can help us to examine the behaviour of a function in the neighbourhood of a critical point. Bear in mind that at such points the function's gradient vanishes.

Considering points in a small region around the critical point $\boldsymbol{a}$ of $f : \mathbb{R}^2 \longrightarrow \mathbb{R}$ we have from Equation (3.1) (ignoring the small terms),

$$f(a+h, b+k) \approx f(a,b) + \frac{1}{2}\left( \frac{\partial^2 f}{\partial x^2}(a,b).h^2 + 2\frac{\partial^2 f}{\partial x \partial y}(a,b).hk + \frac{\partial^2 f}{\partial y^2}(a,b).k^2 \right)$$

$$= f(a,b) + \frac{1}{2}Q(h,k). \qquad \text{— since the gradient term} = 0$$

Here $Q(h,k)$ is called a *quadratic form*. For a general function of $n$ variables which has continuous derivatives of order 2, we can write

$$Q(\boldsymbol{h}) = \boldsymbol{h}^T \cdot H_{\boldsymbol{a}}f \cdot \boldsymbol{h}.$$

For $|\boldsymbol{h}| \ll 1$ the sign of $Q$ determines whether $\boldsymbol{a}$ is a maximum, a minimum, or a saddle point.

Suppose $f : \mathbb{R}^2 \longrightarrow \mathbb{R}$ has continuous derivatives of order no greater than 3 (less than or equal to 3) and $\boldsymbol{a} \in D_f$ is a critical point of $f$.

1) If $Q(h, k)$ is *positive definite*, then $f$ has a local *minimum* value at $\boldsymbol{a}$:
$$Q(h, k) > 0 \text{ for all } 0 \neq |\boldsymbol{h}| \ll 1 \implies f(a + h, b + k) > f(a, b).$$

2) If $Q(h, k)$ is *negative definite*, then $f$ has a local *maximum* value at $\boldsymbol{a}$:
$$Q(h, k) < 0 \text{ for all } 0 \neq |\boldsymbol{h}| \ll 1 \implies f(a + h, b + k) < f(a, b).$$

3) If $Q(h, k)$ is *indefinite*, then $f$ has neither a maximum nor a minimum value at $\boldsymbol{a}$, and $\boldsymbol{a}$ is a *saddle point*: For all $0 \neq |\boldsymbol{h}| \ll 1$,
$$Q(h, k) < 0 \text{ for some } \boldsymbol{h} \implies f(a + h, b + k) > f(a, b),$$
$$Q(h, k) > 0 \text{ for other } \boldsymbol{h} \implies f(a + h, b + k) < f(a, b).$$

4) If $Q(h, k)$ is *positive or negative semi-definite*:
$$Q > 0 \text{ or } Q < 0 \text{ and } Q = 0 \text{ for some } |\boldsymbol{h}| \neq 0,$$
then we cannot say anything.

A summary of this section on critical points and critical point classification appears below, with an aside reviewing the corresponding facts in the case of functions of one variable. The comparison between the 1D and the $n$D cases is quite instructive. The reader should note the similarities at corresponding points of the arguments. Readers will have an opportunity to test their understanding by solving Mastery Checks 3.2–3.5.

In the **one-dimensional case** an investigation into critical points of a function of one variable is summarized as follows.

Consider $y = f(x)$. Critical points are determined from solutions of the zero-derivative equation,
$$\frac{\mathrm{d}f}{\mathrm{d}x}\bigg|_a = 0 \implies x = a$$

As a consequence, we find that

$$\frac{\partial^2 f}{\partial x^2}\bigg|_a > 0 \qquad \Rightarrow \quad \text{minimum}$$

$$\frac{\partial^2 f}{\partial x^2}\bigg|_a < 0 \qquad \Rightarrow \quad \text{maximum}$$

$$\frac{\partial^2 f}{\partial x^2}\bigg|_a = 0 \qquad \Rightarrow \quad \text{a stationary point}$$

In some interval

$$I_r = \{x : 0 < |x - a| < r\}$$

about the critical point, we have the approximation

$$f(x) \approx f(a) + \left.\frac{df}{dx}\right|_a (x - a) + \frac{1}{2}\left.\frac{d^2 f}{dx^2}\right|_a (x - a)^2$$

Hence, for points $x = a + h$ near $x = a$ we have

$$f(a + h) - f(a) \approx \frac{1}{2}\left.\frac{d^2 f}{dx^2}\right|_a h^2$$
$$< 0, \text{ maximum}$$
$$> 0, \text{ minimum}$$
$$= 0, \text{ saddle point.}$$

In the $n$-**dimensional case**, the study of critical points of a function of $n$ variables is very similar.

Consider $z = f(x_1, x_2, \ldots, x_n)$. Critical points are determined by solving

$$\left.\boldsymbol{\nabla} f\right|_a = 0 \Rightarrow \left\{ \begin{array}{c} \dfrac{\partial f}{\partial x_1} = 0 \\ \vdots \\ \dfrac{\partial f}{\partial x_n} = 0 \end{array} \right\} \Rightarrow \boldsymbol{x} = \boldsymbol{a}$$

Having identified a critical point, $\boldsymbol{a}$, we find that

$$\begin{aligned} f(\boldsymbol{x}) &= f(\boldsymbol{a} + \boldsymbol{h}) > f(\boldsymbol{a}) & \forall \boldsymbol{x} \in S_r(\boldsymbol{a}) & \Rightarrow \text{ minimum} \\ f(\boldsymbol{x}) &= f(\boldsymbol{a} + \boldsymbol{h}) < f(\boldsymbol{a}) & \forall \boldsymbol{x} \in S_r(\boldsymbol{a}) & \Rightarrow \text{ maximum} \\ f(\boldsymbol{x}) &\gtrless f(\boldsymbol{a}) & & \Rightarrow \text{ a saddle point} \end{aligned}$$

In some neighbourhood,

$$S_r(\boldsymbol{a}) = \{\boldsymbol{x} : 0 < |\boldsymbol{x} - \boldsymbol{a}| < r\},$$

of the critical point, we have the approximation

$$f(\boldsymbol{x}) \approx f(\boldsymbol{a}) + \left.\boldsymbol{\nabla} f\right|_a \cdot \boldsymbol{h} + \frac{1}{2}\boldsymbol{h}^T \cdot H(\boldsymbol{a}) \cdot \boldsymbol{h}$$

Hence, for points $\boldsymbol{x} = \boldsymbol{a} + \boldsymbol{h}$ near the critical point $\boldsymbol{x} = \boldsymbol{a}$ we have

$$f(\boldsymbol{a} + \boldsymbol{h}) - f(\boldsymbol{a}) \approx \frac{1}{2}\boldsymbol{h}^T \cdot H(\boldsymbol{a}) \cdot \boldsymbol{h} \quad \left(= \frac{1}{2}Q(\boldsymbol{a})\right)$$
$$< 0, \text{ maximum}$$
$$> 0, \text{ minimum}$$
$$= 0, \text{ saddle point.}$$

✒   **Mastery Check 3.2:**

Use the Taylor approximation of order 2 (Section 2.I, Equation (2.7), Page 116) to determine the nature of the critical point for each of the functions $f_1(x, y) = x^2 + y^2 - 2x$ and $f_2(x, y) = x^2 - y^2 - 2x$.

(These functions were the subjects of Example 3.1 and Mastery Check 3.1.)

✒

✒   **Mastery Check 3.3:**

For $z = f(x, y) = \ln(2x^2 + y^2) - 2y$, verify that $\left.\nabla f\right|_{(0,1)} = \mathbf{0}$ at the point $(0, 1)$, but that the function is neither a maximum nor a minimum at this point.

Hint: See Mastery Check 2.25. The graph is in Figure 3.2.

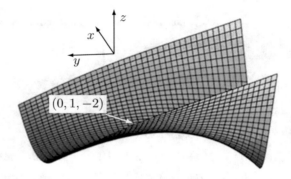

**Figure 3.2** The graph of $z = \ln(2x^2 + y^2) - 2y$.

✒

✒   **Mastery Check 3.4:**

Determine and classify all critical points of the function

$$f(x, y) = x^2 + y^2 + 2e^{xy+1}, \quad D_f = \mathbb{R}^2.$$

✒

✒   **Mastery Check 3.5:**

Determine and classify all critical points of the function

$$f(x, y, z) = x^2y + y^2z + z^2 - 2x, \quad D_f = \mathbb{R}^3.$$

✒

# 3.B   Extreme points: The complete story

According to our discussion in the previous section, a critical point $\boldsymbol{a} \in D_f$ is a point of local maximum if $f(\boldsymbol{a}) \geq f(\boldsymbol{a} + \boldsymbol{h})$ for $|\boldsymbol{h}| \ll 1$ or a point of local minimum if $f(\boldsymbol{a}) \leq f(\boldsymbol{a} + \boldsymbol{h})$ for $|\boldsymbol{h}| \ll 1$.

Such a critical point is also called a point of *relative* maximum or *relative* minimum, respectively.

We contrast these references to *local* and *relative* quantities with the following definitions of *global* quantities.

---

**Definition 3.4**

*A point $\boldsymbol{a} \in D_f$ is called a point of* **absolute** $\begin{cases} maximum \\ minimum \end{cases}$ *if*

$$f(\boldsymbol{a}) \begin{cases} \geq f(\boldsymbol{x}) & \text{for all } \boldsymbol{x} \in D_f \\ \leq f(\boldsymbol{x}) & \text{for all } \boldsymbol{x} \in D_f. \end{cases}$$

---

**Remarks**

∗   The difference between Definition 3.1 and Definition 3.4 lies in the set of points considered. In Definition 3.1 only points in the immediate neighbourhood, $S_r(\boldsymbol{a})$, of $\boldsymbol{a}$ are considered, while in Definition 3.4 all points in the domain, $D_f$, of the function are involved.

∗   Definition 3.4 implies that a critical point, even if a point of local maximum or local minimum, need *not* be a point of *absolute* maximum or minimum.

Earlier we said that critical points are examples of extreme points. However, there are other types of extreme points which are *not* found using the gradient. These are

(i) *singular points of $f$*, points where $\boldsymbol{\nabla} f$ does not exist (Figure 3.3):

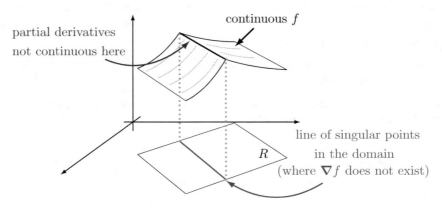

partial derivatives
not continuous here

continuous $f$

line of singular points
in the domain
(where $\boldsymbol{\nabla} f$ does not exist)

$R$

**Figure 3.3**  A function with singular points.

(ii) *boundary points of a restricted region $R \subset D_f$* (Figure 3.4):

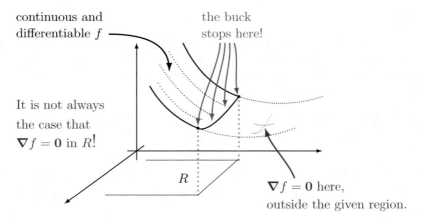

continuous and
differentiable $f$

the buck
stops here!

It is not always
the case that
$\boldsymbol{\nabla} f = \mathbf{0}$ in $R$!

$R$

$\boldsymbol{\nabla} f = \mathbf{0}$ here,
outside the given region.

**Figure 3.4**  A function defined on a restricted region $R \subset D_f$.

Finding the absolute maxima and minima of functions $\mathbb{R}^n \longrightarrow \mathbb{R}$ is part of the science of *optimization*. In general there are at least three categories of optimization problem:

(1) optimizing over a compact domain;

(2) optimizing completely free of restrictions;

(3) optimizing functions under one or more constraints.

We shall study these in turn. We begin with optimizing over compact domains, assuming throughout that the functions involved are differentiable.

## Optimization over compact domains

Recall our definition of a *compact set* (Definition 1.8): a set $\Omega \subseteq \mathbb{R}^n$ is said to be compact if it is *closed* **and** *bounded*.

For any function defined on a region $\Omega \subseteq D_f$ that is compact, we have the following very useful result.

---

**Theorem 3.2**

*A* **continuous** *real-valued function defined on a* **compact** *region,* $\Omega$, *obtains an* **absolute maximum** *and an* **absolute minimum** *value.*

---

A few comments on this theorem are warranted.

Firstly, it is not necessary that the region being considered is the function's entire domain of definition, $D_f$, but it might be. The problem statement will usually specify this. If no region is given then the reader should assume the whole of $D_f$ is implied.

Secondly, by Theorem 1.2, a continuous function defined on a closed and bounded region is necessarily bounded. This means that $|f(\boldsymbol{x})| < K$ for some $K \in \mathbb{R}$ and for all $\boldsymbol{x}$ in that region. This simple result implies that we should expect $f$ to exhibit an absolute minimum and an absolute maximum. In fact, this is the only time we are *guaranteed* that *absolute* maximum and minimum points exist.

The reader should always bear in mind that a *continuous* function is *not* necessarily differentiable everywhere. A consequence of this is that singular points can exist. These should then be inspected separately to any critical points. Naturally, the appealing notion of a closed and finite domain means that the domain boundary (boundary points) need also to be considered separately.

We illustrate Theorem 3.2 in action with the following examples.

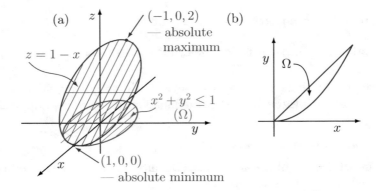

**Figure 3.5**   (a) The graph of $x + z = 1$ in Example 3.3;
(b) The domain $\Omega$ in Example 3.4.

■   **Example 3.3:**

Consider the function

$$f(x, y, z) = x + z = 1,$$

and the region

$$\Omega = \{(x, y) : x^2 + y^2 \le 1\}.$$

both of which are shown in Figure 3.5(a). In this case $\nabla f \neq 0$, but the function still has attained an absolute maximum and an absolute minimum.

■

■   **Example 3.4:**

Consider $f(x, y) = (y - x)e^{x^2 - y}$ for $x^2 \le y \le x$. Where does $f$ achieve its maximum and minimum values?

We note that the domain $\Omega$ set by the above inequalities and shown in Figure 3.5(b) is non-empty provided $0 \le x \le 1$ and is compact, so we are sure to find the extrema. The function is non-singular and we find that

$$\nabla f = \begin{pmatrix} -1 + 2x(y - x) \\ 1 - (y - x) \end{pmatrix} e^{x^2 - y}$$

$$= \mathbf{0} \text{ when } \begin{pmatrix} x \\ y \end{pmatrix} = \begin{pmatrix} 1/2 \\ 3/2 \end{pmatrix},$$

which is outside $\Omega$. So, we need to inspect the boundaries.

On the boundary $y = x$, $f(x, y) = 0$.
On the boundary $y = x^2$, $f(x, y) = x^2 - x$, which has maximum 0 at $x = 0$ and $x = 1$, and minimum $-\frac{1}{4}$ at $x = \frac{1}{2}$, $y = \frac{1}{4}$.
Therefore, the absolute minimum is $f(1/2, 1/4) = -1/4$, and absolute maximum $f(x, x) = 0$.

■

## Optimization free of constraints
In relaxing the condition of compactness, either by allowing the region $R \subseteq D_f$ to be unbounded or bounded but open, there is no longer any guarantee that points of finite maximum or minimum exist. For instance a function might become infinite at one or more points on the boundary of a bounded open set. Consider for example the case

$$f(x, y) = \frac{xy}{\sqrt{1 - x^2 - y^2}}, \qquad R = D_f = \left\{ (x, y) : x^2 + y^2 < 1 \right\}.$$

The magnitude of this otherwise continuous function increases without bound as the independent variables approach the boundary of the unit disc: the function therefore attains neither an absolute maximum nor absolute minimum in $D_f$. In contrast, a continuous function on an unbounded domain may still attain a finite absolute maximum or minimum, as in the case of Mastery Check 3.7:

$$f(x, y) = \frac{4x}{1 + x^2 + y^2}, \qquad R = D_f = \mathbb{R}^2.$$

The function attains both an absolute maximum and an absolute minimum despite an unbounded domain of definition.

So, how does one proceed? We need only to modify the protocol for continuous functions on compact regions. This is the right-hand side of Flowchart 3.1.

We first ascertain if and where there are points on the edge of $D_f$ at which the function $f$ is discontinuous or not defined (that is, it "blows up"), and moreover whether $f$ diverges to positive or negative infinity. This step will immediately answer the question of whether absolute extrema exist at all.

## Flowchart 3.1:   Optimization — how to play the game

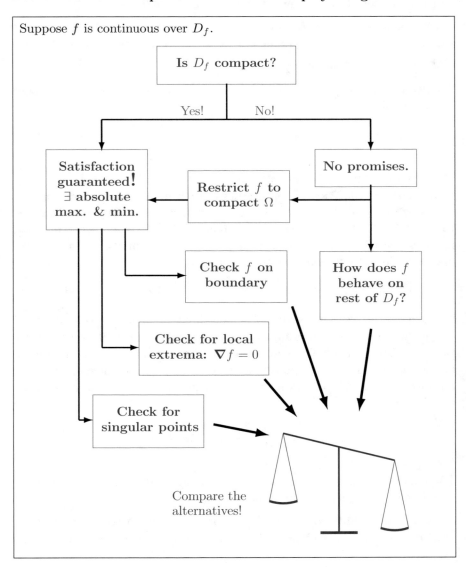

We next work with a convenient closed and bounded subregion of our own choosing over which $f$ is continuous, and whose extent is easily characterized in terms of one or a few parameters; call this subset $\Omega \subset D_f$. Then, confining ourselves to $\Omega$, we proceed as before and identify in $\Omega$ any critical points, points where the function's derivatives don't exist (are singular), and $f$'s behaviour on the boundary of $\Omega$.

Finally, if $D_f$ is unbounded or open we consider the function's behaviour outside of $\Omega$ over the rest of $D_f$.

The results of these three steps are then compared to determine which if any points in $D_f$ are points of absolute maximum and absolute minimum.

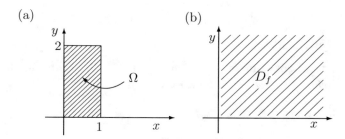

**Figure 3.6**   (a) Bounded domain, $\Omega$, in Mastery Check 3.6;
(b) Open domain, $D_f$, in Mastery Check 3.8.

✍   **Mastery Check 3.6:**

Determine the greatest and least values of

$$f(x,y) = ye^x - xy^2$$

in the region

$$\Omega = \{(x,y) : 0 \le x \le 1,\ 0 \le y \le 2\}$$

shown in Figure 3.6(a).

Hint: check for singular points and interior extreme points, then check boundary points — all of them! Then, and only then, should you plot the graph using MATLAB®.

✍

## ✍ Mastery Check 3.7:

Determine the greatest and least values of

$$f(x,y) = \frac{4x}{1 + x^2 + y^2} \text{ for } (x,y) \in D_f = \mathbb{R}^2.$$

✍

## ✍ Mastery Check 3.8:

Determine the greatest and least values of the function

$$f(x,y) = x + 8y + \frac{1}{xy}$$

on the open domain

$$D_f = \{(x,y) : x > 0, y > 0\}$$

shown in Figure 3.6(b).

Hint: check the behaviour of $f$ for fixed $y$ as $x \to 0, \infty$, and for fixed $x$ as $y \to 0, \infty$, before proceeding to look for extrema.

✍

### Optimization subject to constraints

Function optimization under one or more constraints is a fairly common type of problem. It involves the task of maximizing or minimizing, generally expressed as optimizing, with respect to one or more variables, some quantity under special and restrictive conditions.

Such problems are generally expressed in a common way. For example, we:

| *maximize* | volume | *subject to* | fixed surface area; |
| *minimize* | the physical dimensions of electronic circuitry | *subject to* | limited operating temperature; |
| *optimize* | work & train schedule | *subject to* | a fixed #trains & man hours (40 hr/week). |

Consequently, the generic model structure of such problems (with one constraint) can be expressed as follows:

| *optimize* | $f(\boldsymbol{x})$ | *subject to* | $g(\boldsymbol{x}) = 0$ |

In some applications $f$ is called the *objective* function while $g$ is the *constraint*. In the case where there are more than one constraint, the additional constraints would similarly be expressed as equations such as $h(\boldsymbol{x}) = 0$.

The conceptual picture I like to impart to students is this: Suppose $f$ were a mountain, while $g$ gives rise to a walking track on the mountain side (see Figure 3.7). The constraint $g$ is *not* itself the walking track, but a set of points in the plane of the mountain's base (the domain) that gives rise to the walking track. The absolute unconstrained maximum of $f$ would be the mountain's peak, but the maximum of $f$ subject to constraint $g = 0$ would give only the highest point on the track.

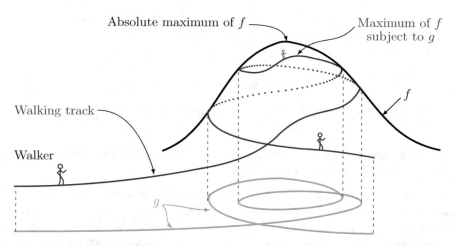

**Figure 3.7**  A constrained optimization problem.

In the following we will assume that both $f$ and $g$ have continuous partial derivatives of order 1 (at least). That is, $f$ and $g$ are of class $C^1$.

We must also assume that *the level set*

$$L = \{\boldsymbol{x} : g(\boldsymbol{x}) = 0\}$$

is a non-empty subset of $D_f$, that is that it lies inside $D_f$. Moreover, we assume that $L$ in the domain of $g$ $(L \subset D_g)$ has a dense set of points in common with $D_f$.

Although the concepts are applicable to any $\mathbb{R}^n$ $(n < \infty)$, it is convenient to restrict the following analysis and discussion to $\mathbb{R}^2$.

Suppose

$$f : \mathbb{R}^2 \longrightarrow \mathbb{R} \quad \text{and} \quad g : \mathbb{R}^2 \longrightarrow \mathbb{R}$$
$$(x, y) \mapsto f(x, y) \qquad (x, y) \mapsto g(x, y)$$

We presume $L \cap D_f \neq \emptyset$ and so we have the following picture, Figure 3.8. Since we consider a function of two variables, $L$ is a curve in the $\mathbb{R}^2$ plane.

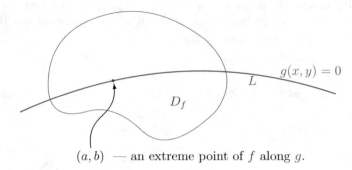

$(a, b)$ — an extreme point of $f$ along $g$.

**Figure 3.8** Domain $D_f$ and the zero level set of $g$, $L$.

Suppose $(a, b)$ is an interior local maximum or local minimum point of $f$ *when* $f$ is restricted to points on $g(x, y) = 0$ in $\mathbb{R}^2$.

As $g$ is a $C^1$ function there exists a continuous and differentiable parametrization of the level set of $g$ in terms of a parameter $t$: $x = \phi(t)$, $y = \psi(t)$, and there exists a value $t_0$ such that $a = \phi(t_0)$, $b = \psi(t_0)$. Moreover, the level curve has a local tangent vector $(\phi'(t), \psi'(t))$ at any point $(\phi(t), \psi(t))$ along the curve. Finally, referring back to Section 2.E and the properties of the gradient, we deduce that $(\phi'(t), \psi'(t))$ is orthogonal to $\nabla g$ at $(a, b)$.

The single-variable function of $t$, $F(t) = f(\phi(t), \psi(t))$, is critical at $t_0$. That is,

$$\left.\frac{\mathrm{d}F}{\mathrm{d}t}\right|_{t_0} = \left.\frac{\partial f}{\partial x}\frac{\mathrm{d}\phi}{\mathrm{d}t}\right|_{t_0} + \left.\frac{\partial f}{\partial y}\frac{\mathrm{d}\psi}{\mathrm{d}t}\right|_{t_0} \qquad \text{— a chain rule application}$$
$$= \nabla f(a, b) \cdot (\phi'(t_0), \psi'(t_0))$$
$$= 0.$$

The last equality implies that the plane vector $\nabla f(a, b)$ and the tangent vector to $g = 0$ at $(a, b)$, $(\phi'(t_0), \psi'(t_0))$, are orthogonal.

Since a tangent vector to the level curve $g = 0$ is orthogonal to the gradient of the function $f$, (Sections 1.F and 2.E), we have substantiated the following theorem.

---

**Theorem 3.3**

*If $(a, b)$ is a point which lies in both $D_f$ and $D_g$, and which is an extreme point of $f : \mathbb{R}^2 \longrightarrow \mathbb{R}$ under the constraint $g(x, y) = 0$, then the vectors $\nabla f(a, b)$ and $\nabla g(a, b)$ are* **parallel.**

---

The reader should note the difference in conditions satisfied by the point $(a, b)$. It is a critical point of "$f$ subject to $g$" (see the gradient equation below), but *not* a critical point of $f$ alone.

We can now argue that if $\nabla g$ is not identically zero, there exists a $\lambda_0 \in \mathbb{R}$ such that

$$\nabla f(a, b) = -\lambda_0 \nabla g(a, b) \quad \Longleftrightarrow \quad \nabla(f + \lambda g)\Big|_{a, b, \lambda_0} = 0,$$

which means that the 3D point $(a, b; \lambda_0)$ is a critical point of the new multi-variable function

$$F(x, y; \lambda) = f(x, y) + \lambda g(x, y).$$

This is called the *Lagrangian function* and $\lambda$ is called the *Lagrange multiplier*.

The pictorial situation showing the relationship between the gradients of $f$ and $g$, and the curves of constant $f$ and of constant $g$, is shown in Figure 3.9.

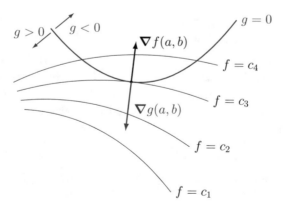

**Figure 3.9**  Level curves and gradient of $f$ relative to the level curve $g = 0$ and $\nabla g$.

**Remarks**

* The critical points of $F(x, y; \lambda)$ are found from the set of equations

$$\left.\begin{array}{l}\dfrac{\partial F}{\partial x} = \dfrac{\partial f}{\partial x} + \lambda \dfrac{\partial g}{\partial x} = 0 \\[2ex] \dfrac{\partial F}{\partial y} = \dfrac{\partial f}{\partial y} + \lambda \dfrac{\partial g}{\partial y} = 0\end{array}\right\} \quad \text{— the condition } \boldsymbol{\nabla} f \| \boldsymbol{\nabla} g \text{ at } (a, b)$$

$$\left.\dfrac{\partial F}{\partial \lambda} = \quad g(x, y) \quad = 0 \right\} \quad \text{— the constraint equation}$$

* What has been done, in effect, is that we have transformed a restricted 2D optimization problem into an *un*restricted 3D optimization problem. We now need to find $a, b,$ *and* $\lambda_0$, to solve the full problem. Note that the actual value of $\lambda_0$ is not often needed, although it can be utilized in the numerical analysis of optimization problems.

* The theory works beautifully for a general function $f : \mathbb{R}^n \longrightarrow \mathbb{R}$ and constraint $g(x_1, \ldots, x_n) = 0$:

$$\boldsymbol{\nabla} f(\boldsymbol{a}) = -\lambda \boldsymbol{\nabla} g(\boldsymbol{a}) \quad \text{if} \quad \boldsymbol{\nabla} g \neq 0.$$

This implies that the higher dimensional point $(\boldsymbol{a}; \lambda)$ is an extreme point of the $(n + 1)$-dimensional space on which the function $F : \mathbb{R}^{n+1} \longrightarrow \mathbb{R}$ is defined.

The theory of the Lagrangian function and Lagrange multiplier generalizes quite naturally to solve problems of optimizing under two or more constraints.

For example, suppose we wish to optimize $f : \mathbb{R}^3 \longrightarrow \mathbb{R}$ subject to the two constraints $g_1(x, y, z) = 0$ and $g_2(x, y, z) = 0$. Then the generalization of the 2D optimization problem would have $\boldsymbol{\nabla} f(\boldsymbol{a})$, $\boldsymbol{\nabla} g_1(\boldsymbol{a})$, and $\boldsymbol{\nabla} g_2(\boldsymbol{a})$, be linearly dependent vectors. This means that there exist points $\boldsymbol{a}$ and scalars $\lambda_0, \mu_0 \in \mathbb{R}$ such that

$$\boldsymbol{\nabla} f(\boldsymbol{a}) + \lambda_0 \boldsymbol{\nabla} g_1(\boldsymbol{a}) + \mu_0 \boldsymbol{\nabla} g_2(\boldsymbol{a}) = 0$$

provided also that $\boldsymbol{\nabla} g_1(\boldsymbol{a})$ and $\boldsymbol{\nabla} g_2(\boldsymbol{a})$ are *not* co-parallel, which can be checked easily by showing that $\boldsymbol{\nabla} g_1(\boldsymbol{a}) \times \boldsymbol{\nabla} g_2(\boldsymbol{a}) \neq 0$.

In turn the linear dependency condition means that the higher dimensional point $(\boldsymbol{a}; \lambda_0, \mu_0)$ is a critical point of

$$F(\boldsymbol{x}; \lambda, \mu) = f(\boldsymbol{x}) + \lambda g_1(\boldsymbol{x}) + \mu g_2(\boldsymbol{x}).$$

Both $\lambda$ and $\mu$ are Lagrange multipliers.

■   **Example 3.5:**

Suppose we wish to find the maximum value of $f(x,y) = e^{-(x^2+y^2)/2}$ subject to the constraint $g(x,y) = y - x^2 + 3 = 0$.

We set up the Lagrangian $F(x,y;\lambda) = e^{-(x^2+y^2)/2} + \lambda(y - x^2 + 3)$. Now

$$\boldsymbol{\nabla} F = \begin{pmatrix} -xe^{-(x^2+y^2)/2} - 2\lambda x \\ -ye^{-(x^2+y^2)/2} + \lambda \\ y - x^2 + 3 \end{pmatrix} = \mathbf{0} \quad \text{when} \quad \begin{pmatrix} x \\ y \\ \lambda \end{pmatrix} = \begin{pmatrix} 0 \\ -3 \\ 3e^{-9/2} \end{pmatrix}.$$

The maximum value of $f$ is $f(0,-3) = e^{-9/2}$.

■

✍   **Mastery Check 3.9:**

Find the maximum and minimum of $f(x,y,z) = x + 2y - 3z$ over the ellipsoid $x^2 + 4y^2 + 9z^2 \le 108$. Try these two methods of solution:

1)   Construct an argument that the maximum and minimum must both be on the surface of the ellipsoid. Then parameterize the ellipsoid in terms of the usual spherical coordinate angles (see Page 37ff) and look for critical points of $f$ in terms of these angles.

2)   Set up the Lagrangian $F = f + \lambda g$ for suitable $g$ and look for critical points of $F$.

Use MATLAB® to draw the graph of the ellipsoid and the maximal level set.

✍

## 3.C   Differentials and error analysis

Definition 2.4 of a differentiable function allows the following interpretation:

$$f(\boldsymbol{x} + \Delta\boldsymbol{x}) - f(\boldsymbol{x}) = \text{grad } f(\boldsymbol{x}) \cdot \Delta\boldsymbol{x} + |\Delta\boldsymbol{x}|\rho(\Delta\boldsymbol{x}),$$

or, more simply, that    $\Delta f(\boldsymbol{x};\Delta\boldsymbol{x}) \approx \text{grad } f(\boldsymbol{x}) \cdot \Delta\boldsymbol{x}$ for $|\Delta\boldsymbol{x}| \ll 1$.

This leads to the idea of constructing a new function of both

$$\boldsymbol{x} = (x_1, x_2, \ldots, x_n) \text{ and } \mathbf{dx} = (dx_1, dx_2, \ldots, dx_n).$$

That is, it is a function of $2n$ variables.

**Definition 3.5**

*The differential of a function $f : \mathbb{R}^n \longrightarrow \mathbb{R}$ is defined as*

$$\mathrm{d}f(\boldsymbol{x}, \mathrm{d}\boldsymbol{x}) = \boldsymbol{\nabla} f(\boldsymbol{x}) \cdot \mathrm{d}\boldsymbol{x}$$
$$= \frac{\partial f}{\partial x_1}(\boldsymbol{x})\mathrm{d}x_1 + \cdots + \frac{\partial f}{\partial x_n}(\boldsymbol{x})\mathrm{d}x_n.$$

The function $\mathrm{d}f$ has the following three features:

(1) It is an approximation to the change in $f$, $\Delta f$, coming from a change $\boldsymbol{x} \to \boldsymbol{x} + \mathrm{d}\boldsymbol{x}$;

(2) it is linear in $\mathrm{d}\boldsymbol{x}$; and

(3) it is a natural tool to use if considering overall error estimates when individual errors ($\mathrm{d}\boldsymbol{x}$) are known.

The last feature identifies the differential's most useful application.

Suppose we have a quantity $f$ whose value depends on many parameters, say $x_1, \ldots, x_n$. Any errors incurred in measuring the $x_i$ result in an error in the quantity $f$.

An estimate of the *maximum error* in $f$ is thus given by

$$|\mathrm{d}f(\Delta\boldsymbol{x})| \leq \left|\frac{\partial f}{\partial x_1}\right|_{\boldsymbol{x}} |\Delta x_1| + \left|\frac{\partial f}{\partial x_2}\right|_{\boldsymbol{x}} |\Delta x_2| + \cdots + \left|\frac{\partial f}{\partial x_n}\right|_{\boldsymbol{x}} |\Delta x_n| \quad (3.2)$$

by the triangle inequality (Section 1.B).

The right-hand side of Equation (3.2) gives the *maximum* possible error only if one knows the *maximum* uncertainty of the individual $x_i$. If one knows the *exact* values of the $\mathrm{d}x_i$ (or $\Delta x_i$) including their signs then we use the differential $\mathrm{d}f(\boldsymbol{x}, \mathrm{d}\boldsymbol{x})$ directly to give an approximate value to the change $\Delta f = f(\boldsymbol{x} + \Delta\boldsymbol{x}) - f(\boldsymbol{x})$.

## 3.D   Method of least squares

In the year 1801 the world of astronomy was excited by the discovery of a new minor planet, Ceres, whose (rough) position in the night sky had been noted a few times before it vanished from view. The young Carl Friedrich Gauss [19, 20] used a method — one he had worked out while still a student — to plot the orbit of the planet, and he was able to tell astronomers where in the sky to search for it. That method is the subject of this section.

In the first example that follows, we imagine the planet's orbit in the night sky to be a straight line which has to be fitted in some optimal fashion to a set of discrete pairs of observations which are subject to errors. In the second example, the method is applied to discrete observations on a supposed planetary orbit. In the third example, the method is extended to continuous domains.

The field of statistics deals with "observations" (measurements) on variables that are known to be subject to random errors. When we observe two or more variables at once, it is often appropriate to ask what is the relationship between these two variables. This question is the basis of the study known as "regression analysis", which is outside the scope of this book. But the core of regression analysis is an application of the differential calculus called the method of least squares, invented independently by Gauss.

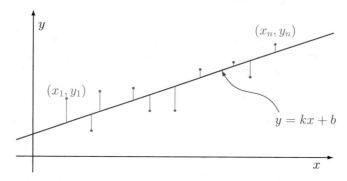

**Figure 3.10**  The line of best fit.

### Fitting a straight line to observations

Suppose we believe that a variable $y$ is in some way dependent on a variable $x$ by the relation $y = kx + b$. In situations such as this it is implicitly assumed that the independent variable is *deterministic*, that is, given, and not subject to error, while the dependent variable is subject to observation or measurement errors.

To find the dependency relationship, we select a sequence of values $\{x_1, x_2, \ldots, x_n\}$ of the independent variable, and measure the corresponding values $\{y_1, y_2, \ldots, y_n\}$ of the dependent variable.

Because observed measurements always have some error associated with them, the observations won't necessarily lie exactly on the straight line, but may fall above or below the line, as in Figure 3.10 above.

**Problem:** How to determine the "line of best fit" $y = k^*x + b^*$ through the "noisy" *discrete* experimental observations. (The values $k^*$ and $b^*$ become estimates of the true parameters $k$ and $b$ in the relation $y = kx + b$ which we believe connects $x$ with $y$.)

**Solution:** We choose the line parameters $k$ and $b$ so that *the sum of the squares of the differences* between the observations and the fitted values is a minimum.

That is, we construct the function $S(k,b) = \sum_{i=1}^{n}(y_i - kx_i - b)^2$ of $k$ and $b$ from the known data and the desired model and seek its minimum to give us the optimal $k$ and $b$ values. We apply the techniques of the preceding chapter to get the critical points of $S$ by solving the two equations

$$\frac{\partial S}{\partial k} = 0 \quad \text{and} \quad \frac{\partial S}{\partial b} = 0.$$

Because of its form $S$ has no upper bound but does have a lower bound. Moreover, there will be only one critical point which will be the $(k, b)$ point for which $S$ has its minimum value. In fact, with the above explicit expression for $S$ we get the equations

$$\sum_{i=1}^{n} 2(y_i - k^*x_i - b^*)(-x_i) = 0 \quad \text{and} \quad \sum_{i=1}^{n} 2(y_i - k^*x_i - b^*)(-1) = 0.$$

That is, $\sum_{i=1}^{n} x_i y_i - k^* \sum_{i=1}^{n} x_i^2 - b^* \sum_{i=1}^{n} x_i = 0$ and $\sum_{i=1}^{n} y_i - k^* \sum_{i=1}^{n} x_i - nb^* = 0.$

This pair of equations leads to the $2 \times 2$ matrix equation

$$\begin{pmatrix} \sum x_i^2 & \sum x_i \\ \sum x_i & n \end{pmatrix}\begin{pmatrix} k^* \\ b^* \end{pmatrix} = \begin{pmatrix} \sum x_i y_i \\ \sum y_i \end{pmatrix},$$

which can be solved to give the optimal $k^*$ and $b^*$.

## ✍ Mastery Check 3.10:

Invert this matrix equation to obtain explicit estimates of $k^*$ and $b^*$.

Hint: First define symbols $\bar{x} = \sum x_i/n$, etc.

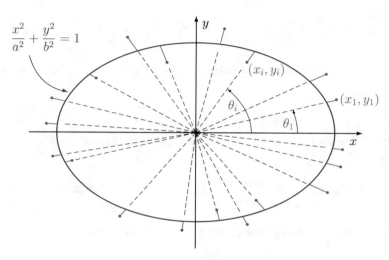

**Figure 3.11**   The ellipse of best fit.

### Fitting a conic to observations

There is nothing unique about fitting straight lines: observations can follow other functional forms depending on the problem being considered.

Suppose, for example, the observations $\{(x_i, y_i)\}$ are scattered in the shape of an ellipse with the origin as centre. (Remember the case of the minor planet.) We want to determine that ellipse,

$$\frac{x^2}{a^2} + \frac{y^2}{b^2} = 1,$$

with optimal $a$ and $b$ that best represent the observations.

It is convenient at this point to convert to polar coordinates (Page 25)

$$(x_i, y_i) \longrightarrow (r_i, \theta_i) : \begin{cases} x_i = r_i \cos\theta_i, \\ y_i = r_i \sin\theta_i, \end{cases} i = 1, 2, \ldots, n.$$

Choose as error function $E$ the sum of squares of the distances of the observation pairs $(x_i, y_i)$ from the corresponding points $\big(x(\theta_i), y(\theta_i)\big)$ on the ellipse along the rays at the given $\theta_i$, as in Figure 3.11.

What we are doing from here on is assuming that the angles $\theta_i$ have no associated errors, but that the radial distances $r_i$ *are* subject to error. Analogous to the straight line example we form the sum of squares of differences

$$E = \sum_{i=1}^{n} \left[ \left(x(\theta_i) - x_i\right)^2 + \left(y(\theta_i) - y_i\right)^2 \right]$$

$$= \sum_{i=1}^{n} \left[ (a - r_i)^2 \cos^2 \theta_i + (b - r_i)^2 \sin^2 \theta_i \right]$$

As before, look for the critical points of $E(a, b)$ with respect to $a$ and $b$.

$$\frac{\partial E}{\partial a} = 0 \quad \text{and} \quad \frac{\partial E}{\partial b} = 0.$$

The $2 \times 2$ system of equations can be solved to give the optimal ellipse parameters

$$a^* = \frac{\sum_{i=1}^{n} r_i \cos^2 \theta_i}{\sum_{i=1}^{n} \cos^2 \theta_i} \quad \text{and} \quad b^* = \frac{\sum_{i=1}^{n} r_i \sin^2 \theta_i}{\sum_{i=1}^{n} \sin^2 \theta_i}.$$

Try to confirm these expressions for your own peace of mind!

## Least-squares method and function approximations

The least-squares method admits a continuous version in which the discrete sum is replaced with an integral, and the discrete observations are replaced with a continuously varying function. In truth, the function need not be continuous, only integrable. However, for this introduction we'll stay with the more restrictive but less complicated case.

Suppose $f : \mathbb{R} \longrightarrow \mathbb{R}$ is a continuous function on some interval $a \leq x \leq b$. The problem that is now posed is how do we "best" approximate $f$ with an "approximating" function $g(x; \lambda_1, \ldots, \lambda_n)$, where $\{\lambda_i\}_{i=1}^{n}$ is a set of $n$ parameters?

In solving this problem we first note that "best" will depend on how this is measured. That is, this will depend on the choice of *distance function*. (See [9] and [10] for a more complete discussion.) Second, we note that the choice of the approximating function $g$ is critical.

In the theme of least squares, the distance function most often considered is

$$I = \int_{a}^{b} \left( f(x) - g(x; \lambda_1, \ldots, \lambda_n) \right)^2 \, \mathrm{d}x$$

which mathematicians refer to as the $L_2$ distance function.

The choice of the approximating function $g$ is usually (but not always) a linear function of the $\lambda_i$. The general structure is

$$g(x; \lambda_1, \ldots, \lambda_n) = \sum_{i=1}^{n} \lambda_i \phi_i(x),$$

where the functions $\phi_i(x)$ are chosen to satisfy some criteria specific to the problem being considered.

The best approximation to $f$ is then the choice of the $\lambda_i$ that minimize the distance function. Thus, we look for critical points of $I$:

$$\left.\begin{aligned} \frac{\partial I}{\partial \lambda_1} &= 0 \\ \vdots \quad &\quad \vdots \\ \frac{\partial I}{\partial \lambda_n} &= 0 \end{aligned}\right\} \quad n \text{ equations in } n \text{ unknowns} \quad \Longrightarrow \quad \lambda_i^*, i = 1, \ldots, n.$$

Some common and useful, but not exclusive, choices of $g$ (that is, the $\phi_i$) are:

$$g(x; \lambda_0, \ldots, \lambda_n) = \lambda_0 + \lambda_1 x + \cdots + \lambda_n x^n \quad \text{—polynomial approximation}$$

$$g(x; \lambda_1, \ldots, \lambda_n) = \sum_{k=1}^{n} \lambda_n \sin kx \quad \text{—trigonometric (sine) approximation}$$

Notice that in these cases the $\lambda_i$ appear as linear coefficients.

✍ **Mastery Check 3.11:**

Find constants $a, b, c \in \mathbb{R}$ which minimize the integral

$$J = \int_{-1}^{1} \left( x - a\sin(\pi x) - b\sin(2\pi x) - c\sin(3\pi x) \right)^2 \mathrm{d}x$$

$$= \int_{-1}^{1} \left( x - g(x; a, b, c) \right)^2 \mathrm{d}x.$$

Plot $f(x) = x$ and $g(x)$ over the interval $[-1, 1]$.

✍

# 3.E Partial derivatives in equations: Partial differential equations

Although we treat the topic of partial differential equations (PDEs) as an application of differential calculus, the body of knowledge is immense and can (and sometimes does) easily cover several thick volumes of dedicated texts. Our coverage keeps with the idea of giving only some basic practical information that students of engineering and science will find beneficial in their undergraduate studies, and possibly later.

Although PDEs—which are simply scalar-valued equations that describe fixed relationships between the various partial derivatives of a scalar function—come in all shapes and sizes, the type that has been most thoroughly studied and on which we will focus is the *second-order, linear PDE*. The following definition features a scalar function $u$ of two independent variables. However, second-order linear PDEs may depend on more than two variables.

---

**Definition 3.6**

*A second-order, linear PDE with constant coefficients satisfied by a function $u \in C^2(\mathbb{R}^2)$ on a domain $D_u \subset \mathbb{R}^2$ is an equation of the form*

$$Lu = Au_{xx} + Bu_{xy} + Cu_{yy} + Du_x + Eu_y + Fu + G = 0,$$

*where the constant real numbers $A, B, C, D, E, F$ satisfy the condition $A^2 + B^2 + C^2 \neq 0$, and $G(x,y)$ is a real-valued function defined on $D_u$.*

---

Even in this restrictive case there is a quite a lot known. Unfortunately, the following remarks and other comments made in this section do not do justice to the discipline or to the effort that has gone into accumulating all that is known. Nevertheless, it is hoped that the reader will be somewhat better oriented with the subject with the salient terminology and information provided here.

**Remarks**

* In this section subscripts $x$, $y$, and $t$ on the dependent variable $u$ will denote partial derivatives w.r.t. those variables.

* The "$L$" as used here and in many texts will denote a "linear partial differential operator" with the property that $L(u + v) = Lu + Lv$.

* "Second order" refers to the highest *order* of partial derivative appearing in an equation.
  — a *third order* PDE would have one or more of $u_{xxx}, u_{xxy}, \ldots, u_{yyy}$.

* "Linear" refers to PDEs containing terms which are *at most* proportional to $u$ or one of its derivatives. — $G(x,y)$ *does not* involve $u$.

* The term "PDE" means an equation relating a function to its partial derivatives, or relating partial derivatives to one another.

* The function $u$ is called a *solution* of the PDE.

* The PDE relationship applies locally in an open domain $D_u$ but not on the boundary $\partial D_u$.

* A *time-dependent* PDE generalizes Definition 3.6 to include an explicit dependence on the time variable $t$ through additional terms involving partial derivatives of $u$ w.r.t. $t$.

---

**Definition 3.7**
*The quantity $B^2 - 4AC$ is called the discriminant of the PDE.*

---

The reader may speculate that this discriminant is somehow connected with the discriminant appearing in the solution of a quadratic equation. The connection *is* there and *is* important but would take us too far afield to explain what it is about. The curious reader might refer to the bibliography [11–13, 15] for more information. However, for our purposes it is sufficient to just state without proof some important facts.

Firstly, the value of the discriminant (more significantly its sign) determines the type or nature of the PDE. A PDE is said to be of

(a) hyperbolic type $\iff$ $B^2 - 4AC > 0$;

(b) elliptic type $\iff$ $B^2 - 4AC < 0$;

(c) parabolic type $\iff$ $B^2 - 4AC = 0$.

These types of PDE exhibit distinctively different behaviour with regard to their properties and conditions. So too do their solutions.

**Summary of some relevant facts**

∗ Linearity

- Adding any number of possible solutions leads to a new solution. Solutions $u_1(x, y)$, $u_2(x, y)$, and constants $\alpha$ and $\beta$, can be combined to give a new solution: $\alpha u_1(x, y) + \beta u_2(x, y)$

- The solutions are not unique, since any scalar multiple of a solution is also a solution.

- To get uniqueness we impose additional conditions, which are used to specify the scalar multipliers needed. These conditions are classed as either *initial* conditions or *boundary* conditions.

∗ Initial conditions
Initial conditions are needed for the solution of time-dependent PDEs and inform how the solution begins. They are of the form:

- "displacement": $u(\cdot, t = 0) = u_0(\cdot)$, with $u_0(\cdot)$ specified;

- "velocity": $u_t(\cdot, t = 0) = v_0(\cdot)$, with $v_0(\cdot)$ specified;

where the "·" indicates that other independent variables apply.

∗ Boundary conditions
Boundary conditions on the domain boundary $\partial D_u$ relate to spatial-dependent PDEs.

These are of three forms:

- Dirichlet conditions
  $u(\boldsymbol{r} = \boldsymbol{r}_b, t) = U_b(\boldsymbol{r}_b, t)$, with $U_b$ specified at $\boldsymbol{r}_b \in \partial D_u$.

- Neumann conditions (Figure 3.12)
  Let $\boldsymbol{N}(\boldsymbol{r}_b)$ be the outward normal to the boundary $\partial D_u$. Then the Neumann condition fixes the normal component of the solution gradient $\boldsymbol{\nabla} u(\boldsymbol{r} = \boldsymbol{r}_b, t) \cdot \boldsymbol{N}(\boldsymbol{r}_b) = V_b$, with $V_b$ specified at $\boldsymbol{r}_b \in \partial D_u$.

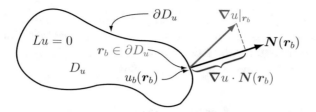

**Figure 3.12**  Neumann condition on the boundary of domain $D_u$.

- Cauchy conditions
  These specify both the solution and the normal component of the solution gradient at the same point.

  $$\begin{cases} u(\boldsymbol{r} = \boldsymbol{r}_b, t) = U_b(\boldsymbol{r}_b, t), \\ \boldsymbol{\nabla} u(\boldsymbol{r} = \boldsymbol{r}_b, t) \cdot \boldsymbol{N}(\boldsymbol{r}_b) = V_b(\boldsymbol{r}_b, t), \end{cases}$$

  with $U_b$ and $V_b$ specified at $\boldsymbol{r}_b \in \partial D_u$.

* Mixed conditions
  These are needed for solutions of PDEs that depend on both time and space in bounded spatial domains.

  For example,    $$\begin{cases} u(\boldsymbol{r}, t = 0) = u_0(\boldsymbol{r}), \\ u(\boldsymbol{r} = \boldsymbol{r}_b, t) = U_b(\boldsymbol{r}_b), \\ \boldsymbol{\nabla} u(\boldsymbol{r} = \boldsymbol{r}_b, t) \cdot \boldsymbol{N}(\boldsymbol{r}_b) = V_b(\boldsymbol{r}_b). \end{cases}$$

  ($U_b, V_b$ are not time-dependent. Contrast this set with the boundary condition set on Page 162, Equation (3.5).)

* Rule of thumb for applying these conditions

  - We need as many initial conditions as the order of the highest order time derivative;

  - We need as many boundary conditions as the order of the highest order spatial derivative in *each independent* variable.

* Types of PDE

  - The discriminant (Definition 3.7) indicates the type of PDE under consideration. In turn, the type dictates what additional conditions are needed and allowed in order to determine a unique solution to the problem.

- It is easily shown [12] that any second-order, linear PDE can be reduced to one of three forms in the respective classes of *hyperbolic*, *elliptic*, and *parabolic* PDE. We illustrate these forms by means of the most common example from each class.

## Classic equations of mathematical physics

Heeding the last bullet point, we focus attention on three classic equations of mathematical physics: the Laplace equation (*elliptic*), the diffusion equation (*parabolic*) and the wave equation (*hyperbolic*). The applications in which they arise are mentioned in their respective sections below. Suffice to say that their solutions play central roles in describing a wide range of physical phenomena.

### The Laplace equation

The Laplace equation is perhaps the PDE with the longest history, dating back to the late 18$^{\text{th}}$ century when it was introduced by Pierre-Simon Laplace [17] in the context of astronomy. Apart from being the equation satisfied by a free-space gravitational potential, the PDE also arises in the context of the steady dynamics of an incompressible fluid, and is satisfied by the electrostatic potential in electrostatic theory and the velocity potential in linear acoustic theory [13, 15]. Despite their different physical origins, the mathematical properties of the potential functions are the same.

---

**Definition 3.8**
*Suppose $u \in C^2(\mathbb{R}^3)$. The equation*

$$u_{xx} + u_{yy} + u_{zz} = 0 \tag{3.3}$$

*satisfied by $u$ in an open domain $D_u \subset \mathbb{R}^3$ is called the (three-dimensional)* **Laplace equation (potential equation).**

---

**Remarks**

* The Laplace equation is an example of an *elliptic* PDE.

* The appropriate boundary conditions to use to establish a unique solution of Laplace's equation are of Dirichlet type (Page 154).

> **Definition 3.9**
> ***The 2D Dirichlet Problem***
> *Let $D_u$ be a simply-connected bounded region in $\mathbb{R}^2$ and let $\partial D_u$ be its boundary. Let $g(x, y)$ be a continuous function on $\partial D_u$. Then the Dirichlet problem is to find the (unique) function $u = f(x, y)$ such that*
>
> (a) $u$ is defined and continuous on $\bar{D}_u$,
>
> (b) $u$ satisfies the 2D Laplace equation, $u_{xx} + u_{yy} = 0 \ \forall \ (x, y) \in D_u$,
>
> (c) $u = g(x, y) \ \forall \ (x, y) \in \partial D_u$.

* There exists a branch of pure mathematics called *harmonic analysis* that specializes on properties and behaviour of solutions of the 2D Laplace equation.

> **Definition 3.10**
> *A function $u = f(x, y) \in C^2(\mathbb{R}^2)$ that is a solution of the 2D Laplace equation (3.3) is called a* **harmonic function**.

* Equation (3.3) is sometimes abbreviated $\quad \Delta u \equiv \nabla^2 u \equiv \boldsymbol{\nabla} \cdot \boldsymbol{\nabla} u = 0$, where $\boldsymbol{\nabla} = \left(\dfrac{\partial}{\partial x}, \dfrac{\partial}{\partial y}, \dfrac{\partial}{\partial z}\right)$ is the gradient operator, and $\Delta \equiv \nabla^2 \equiv \boldsymbol{\nabla} \cdot \boldsymbol{\nabla}$ — called the *Laplace operator* or *Laplacian operator*.

* All terms appearing in the Laplace equation involve the unknown function $u$. It is therefore said to be a *homogeneous* PDE. If a term not involving $u$ were present it would then be an example of an *inhomogeneous* equation.

> **Definition 3.11**
> *Suppose $u \in C^2(\mathbb{R}^3)$ and $g \in C(\mathbb{R}^3)$. The equation*
> $$\Delta u = u_{xx} + u_{yy} + u_{zz} = g(x, y, z)$$
> *satisfied by $u$ in an open domain $D_u \cap D_g \subset \mathbb{R}^3$ is called a* **Poisson equation** *(inhomogeneous version of the Laplace equation).*

* The Laplacian operator (appearing in 3D diffusion, potential, and wave problems) can be expressed in different forms depending on the problem

geometry. The most common and the most studied forms are those shown in Figure 3.13 (see Section 1.D).

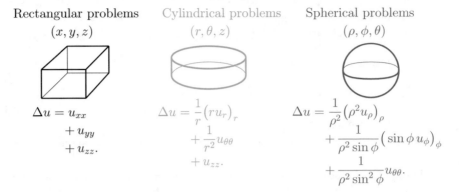

Rectangular problems $(x, y, z)$

$$\Delta u = u_{xx}$$
$$+ u_{yy}$$
$$+ u_{zz}.$$

Cylindrical problems $(r, \theta, z)$

$$\Delta u = \frac{1}{r}\left(r u_r\right)_r$$
$$+ \frac{1}{r^2} u_{\theta\theta}$$
$$+ u_{zz}.$$

Spherical problems $(\rho, \phi, \theta)$

$$\Delta u = \frac{1}{\rho^2}\left(\rho^2 u_\rho\right)_\rho$$
$$+ \frac{1}{\rho^2 \sin \phi}\left(\sin \phi \, u_\phi\right)_\phi$$
$$+ \frac{1}{\rho^2 \sin^2 \phi} u_{\theta\theta}.$$

**Figure 3.13**  The Laplacian expressed in different curvilinear coordinates.

In Figure 3.13 we use the notation introduced incidentally in the comment immediately following Definition 2.3, where subscripts denote partial differentiation with respect to the variable featured in the subscript. For example,

$$\left(r u_r\right)_r \equiv \frac{\partial}{\partial r}\left(r \frac{\partial u}{\partial r}\right).$$

**The diffusion equation**

The diffusion equation is also commonly referred to as the equation of conduction of heat. The equation is satisfied by a function describing the temporal and spatial development of temperature in uniform medium. In the early 1800s Jean-Baptiste Joseph Fourier [17] provided the first in-depth study of this equation in the context of heat transfer, and of its solution by an innovative (for that time) solution method [13, 15]. The equation, with the same form and therefore with the same mathematical properties of its solution, is also satisfied by a function describing the concentration of material diffusing by random processes. The diffusion equation also arises in problems as diverse as radiative transfer, insect migration and the spread of infection.

**Definition 3.12**

*Suppose $u \in C^2(\mathbb{R}^3 \times [0, \infty))$. The equation*

$$\frac{\partial u}{\partial t} - k\left(\frac{\partial^2 u}{\partial x^2} + \frac{\partial^2 u}{\partial y^2} + \frac{\partial^2 u}{\partial z^2}\right) = 0$$

*satisfied by $u$ in the domain $D_u \subset \mathbb{R}^3 \times [0, \infty)$ is called the (three-dimensional)* **diffusion equation** *(***heat equation***).*

**Remarks**

* The diffusion equation is an example of a *parabolic* PDE.

* The appropriate supplementary conditions for the diffusion equation are of mixed type, with one initial condition and Dirichlet or Neumann boundary conditions.

* When $u$ is the *temperature* at point $\boldsymbol{r}$ at time $t$ the diffusion equation is called the *heat equation*.

* The constant $k$ is called the thermal *diffusivity*: $k = \dfrac{K}{s\rho}$, where $K$ is thermal *conductivity*, $\rho$ is *mass* density, and $s$ is *specific heat*.

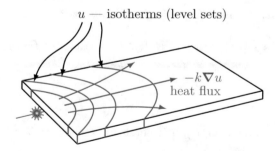

**Figure 3.14** Illustration of heat diffusion in a slab.

The observant reader will have noticed our depiction here of the local heat flux being orthogonal to the isotherms. The gradient operator has that property, and moreover points in the direction of increasing scalar (isotherm) values. The negative sign inherent in the heat flux definition, shown in Figure 3.14, reverses that direction.

* The 3D diffusion (heat) equation can be abbreviated

$$\frac{\partial u}{\partial t} = k\boldsymbol{\nabla} \cdot \boldsymbol{\nabla} u = k\Delta u$$

($\Delta$ — the Laplacian operator. )

* In the limit $t \to \infty$, $\dfrac{\partial u}{\partial t} \to 0$, and $u(\boldsymbol{r}, t \to \infty) \to u_\infty(\boldsymbol{r})$,

the steady-state solution of the Laplace equation: $\Delta u_\infty = 0$. Thus, the Laplace equation is the temporal limit of the diffusion equation.

### The boundary-value problem for the diffusion equation

For some $b > 0$ consider $\bar{R}$ to be the closed set $\bar{R} = \{(x,t) : 0 \leq x \leq \pi, \ 0 \leq t \leq b\}$ with boundary $\partial R = \Gamma_1 \cup \Gamma_2 \cup \Gamma_3 \cup \Gamma_4$ where

$$\Gamma_1 = \{(0,t) : 0 \leq t \leq b\}, \ \Gamma_2 = \{(x,0) : 0 \leq x \leq \pi\},$$
$$\Gamma_3 = \{(\pi,t) : 0 \leq t \leq b\}, \ \Gamma_4 = \{(x,b) : 0 \leq x \leq \pi\}.$$

---

**Definition 3.13**

*The boundary-value problem for the 1D heat-conduction problem is to find a function $u = f(x,t)$ such that*

*(a) $f$ is continuous on $\bar{R}$,*

*(b) $f$ satisfies the heat (diffusion) equation in $R$,*

*(c) $f$ satisfies $f(0,t) = g_1$ on $\Gamma_1$; $f(\pi,t) = g_3$ on $\Gamma_3$; $f(x,0) = g_2$ on $\Gamma_2$.*

---

### Remarks

* Note that there is no explicit condition to be satisfied on $\Gamma_4$!

* The upper bound $b$ could be taken as large as desired. The only restriction is that $b > 0$. That is, we *solve forward in time*. In fact, we often take $b = \infty$ so the time variable $t \in [0, \infty)$.

* The conditions to be applied on $\Gamma_1$ and $\Gamma_3$ are *boundary conditions*, while the condition on $\Gamma_2$ is an *initial condition*.

And, in terms of the time variable $t$ and space variable $x$,

* The boundary conditions $u(0,t) = g_1(t)$ and $u(\pi,t) = g_3(t)$ describe the case of prescribed end temperatures, which may be time-dependent.

* The boundary conditions on $\Gamma_1$ and $\Gamma_3$ may instead be $u_x(0, t) = h_1(t)$ and $u_x(\pi, t) = h_3(t)$ which describe the case of *prescribed end heat fluxes*: rates at which heat is conducted across the ends.

  If $h_1 = h_3 = 0$, the ends are *insulating*. That is, there is no heat conduction.

## The wave equation

The wave equation is the third classic equation of mathematical physics we discuss. As the name suggests it is an equation governing wave-like phenomena; not simply propagation, but oscillatory motion. Its discovery is associated with the names of the 18$^{\text{th}}$-century mathematicians Jean-Baptiste d'Alembert (1D version) and Leonard Euler (3D version) [17]. The equation governs the linear dynamics of water waves, sound waves in linear acoustics, linear elastic vibrations in solid mechanics, and light waves in electromagnetism [13, 15].

---

**Definition 3.14**

*Suppose $u \in C^2(\mathbb{R}^3 \times \mathbb{R})$. The equation*

$$u_{tt} - c^2(u_{xx} + u_{yy} + u_{zz}) = G(x, y, z, t) \qquad (3.4)$$

*satisfied by $u$ in an open domain $D_u \subset \mathbb{R}^3 \times \mathbb{R}$ is called the (three-dimensional)* **wave equation**. *Here, $c$ is called the* wave speed, *and $G$ is defined on $D_u$.*

---

## Remarks

* The wave equation is an example of a *hyperbolic* PDE.

* The appropriate supplementary conditions for a unique solution of the wave equation are either of Cauchy type for unbounded domains or of mixed type for bounded domains.

* Unlike the diffusion equation, the wave equation can be solved forwards or backwards in time.

* If $G(x, y, z, t) \equiv 0$ then Equation (3.4) is the *homogeneous* wave equation.

* The wave speed $c$ is also called the phase speed of propagation of information.

∗ The general form of Equation (3.4) can be abbreviated

$$\frac{\partial^2 u}{\partial t^2} - c^2 \boldsymbol{\nabla} \cdot \boldsymbol{\nabla} u = \frac{\partial^2 u}{\partial t^2} - c^2 \Delta u = G(x, y, z, t)$$

($\Delta$ — the Laplacian operator. $\uparrow$ )

## A general-purpose method of solution: Separation of variables

The method we shall now describe is based on the fact that the linear PDEs just described are *separable* in a number of coordinate systems. This means that their solutions can be expressed in terms of factors involving only one variable. A necessary assumption is that one of the coordinates of the system of choice is constant over a surface on which boundary conditions are prescribed. The method is actually applicable to a wider variety of linear PDEs than the ones that are here highlighted, defined on bounded or semi-bounded domains. Consequently, all students should be familiar with this method of solution [12, 13, 15]. We illustrate the approach by applying it to a simple problem involving 1D wave propagation.

Consider the following *mixed boundary-value problem* on the space-time, semi-infinite strip $D_u = [a, b] \times [0, \infty)$ shown in Figure 3.15. Note that this problem involves one space dimension $x$ and one time dimension $t$.

---

**Definition 3.15**
**The 1D wave problem.** *Find a function $u = f(x, t)$ on $D_u \subset \mathbb{R} \times [0, \infty)$ such that for functions $g_1$ and $g_2$ defined on $[a, b]$ and continuous functions $h_1$ and $h_2$ on $[0, \infty)$, $u$ satisfies the homogeneous wave equation*

$$u_{tt} - c^2 u_{xx} = 0,$$

$$\left.\begin{array}{l} with\ \textbf{initial conditions} \left\{ \begin{array}{l} f(x,0) = g_1(x) \\ f_t(x,0) = g_2(x) \end{array} \right\} a \le x \le b \\[2ex] and\ \textbf{boundary conditions} \left\{ \begin{array}{l} f(a,t) = h_1(t) \\ f(b,t) = h_2(t) \end{array} \right\} t \ge 0. \end{array}\right\} \quad (3.5)$$

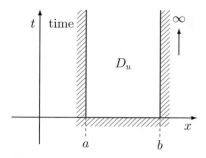

**Figure 3.15**   Domain $D_u$ of the 1D wave problem.

## Remarks

* Equation (3.5) is a quite general form of a mixed b.v.p. for the wave equation in one space variable, but not the most general (see Supplementary problem 21).

* The smoothness (that is, degree of differentiability) of the boundary data, $h_1$ and $h_2$, and the initial data, $g_1$ and $g_2$, determine the smoothness of the solution $u = f(x,t)$.

To illustrate the separation of variables method we consider the special case:

$$(h_1, h_2, g_1, g_2) = (0, 0, g, 0).$$

Similar analyses apply in the other special cases:   $(0, 0, 0, g)$, $(0, h, 0, 0)$,....

For the most general case of $h_1$, $h_2$, $g_1$, and $g_2$ all nonzero, we appeal to the *principle of superposition* (defined shortly). For the moment, we consider the simpler mixed b.v.p.

$$\left.\begin{array}{r}
u_{tt} - c^2 u_{xx} = 0, \\
f(x,0) = g(x), \, 0 \le x \le L \\
f_t(x,0) = 0, \quad 0 \le x \le L \\
f(0,t) = f(L,t) = 0, \quad t \ge 0
\end{array}\right\} \tag{3.6}$$

## Remarks

* Equation (3.6) is the mathematical model of a vibrating string of length $L$ and density $\rho$ under tension $\tau$, fixed at its ends and subject to small displacements elsewhere along its length (Figure 3.16).

* $c = \sqrt{\dfrac{\tau}{\rho}}$; the *wave speed* is determined by the string properties.

* If $g \equiv 0$ then the unique solution would be $f \equiv 0$.

          — a typical way of proving uniqueness

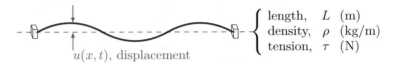

$$\begin{cases} \text{length,} & L \ \text{(m)} \\ \text{density,} & \rho \ \text{(kg/m)} \\ \text{tension,} & \tau \ \text{(N)} \end{cases}$$

$u(x, t)$, displacement

**Figure 3.16** Schematic of a vibrating string.

For the vibrating string problem, $u$ ($|u| \ll L$) is the local dynamic displacement of the stretched string at position $x$ at time $t$. One sets the string vibrating either by "plucking" or "striking", or any linear combination of these. In our example we have chosen the first means. Suppose the string is pulled aside at $t = 0$ a distance $h$ from equilibrium at $x = b$ (Figure 3.17) and released. This condition defines the function $g(x)$.

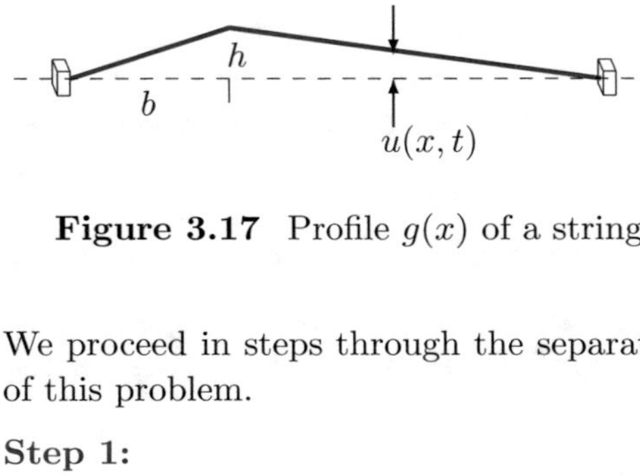

$$g(x) = \begin{cases} \dfrac{xh}{b} & \text{for} \ \ x \in [0, b] \\[2mm] \dfrac{(L - x)h}{L - b} & \text{for} \ \ x \in [b, L]. \end{cases}$$

$u(x, t)$

**Figure 3.17** Profile $g(x)$ of a string stretched a distance $h$ at $x = b$.

We proceed in steps through the separation of variables method of solution of this problem.

**Step 1:**

The separation of variables method always begins by assuming a nontrivial solution in the form of a product of functions of the independent variables. In this case we assume the form

$$f(x, t) = X(x)T(t).$$

That is, the solution is a product of a function of $x$ only ($X$) and a function of $t$ only ($T$).

Substitution into Equation (3.6a) gives

$$X(x)\frac{\mathrm{d}^2 T}{\mathrm{d}t^2} - c^2 T(t)\frac{\mathrm{d}^2 X}{\mathrm{d}x^2} = 0 \tag{3.7}$$

and assuming $X(x) \neq 0$ and $T(t) \neq 0$, Equation (3.7) implies

$$\frac{1}{X(x)}\frac{\mathrm{d}^2 X}{\mathrm{d}x^2} = \frac{1}{c^2}\frac{1}{T(t)}\frac{\mathrm{d}^2 T}{\mathrm{d}t^2} = \mu$$

$$\left\{\begin{array}{c}\text{function}\\\text{of }x-only\end{array}\right\} = \left\{\begin{array}{c}\text{function}\\\text{of }t\text{-only}\end{array}\right\} = \{\text{constant}\}$$

(3.8)

The most critical element of the separation of variables method is the fact that the two expressions on the left of Equation (3.8) can be equal *only* if both equal the same constant! This allows the separation to occur. Equation (3.8) implies two equations

$$\left.\begin{array}{c}\dfrac{\mathrm{d}^2 X}{\mathrm{d}x^2} = \mu X\\[2mm]\dfrac{\mathrm{d}^2 T}{\mathrm{d}t^2} = \mu c^2 T\end{array}\right\}\left(\begin{array}{c}\text{Variables }x\text{ and }t\\\text{have been separated}\end{array}\right)$$

(3.9)

and we have the critical simplifying step:

*One PDE in two variables becomes two ordinary differential equations (ODEs) in single variables!*

This separation is not always possible with other types of problems. Moreover, with some boundaries and some boundary conditions such a product form may not be possible either, even if the PDE allows it.

From this point on we need only to consider the solutions of the ordinary differential equations.

## Step 2:

Consider possible values of $\mu$. The boundary conditions (Equation (3.6d)) imply that

$$X(0) = X(L) = 0.$$

(3.10)

These are necessary for a unique solution.

**Case 1:** $\mu = 0$
Equation (3.9a) implies that $X'' = 0$, which is true if and only if $X(x) = c_1 x + c_2$. But the boundary conditions in Equation (3.10) imply $c_1 = c_2 = 0$. That is, $X(x) \equiv 0$

— the trivial solution.

**Case 2:** $\mu > 0$, that is, $\mu = k^2 > 0$.
Equation (3.9a) implies that $X'' - k^2 X = 0 \iff X(x) = c_1 e^{kx} + c_2 e^{-kx}$.

But the boundary conditions in Equation (3.10) imply

$$x = 0: \quad c_1 + c_2 \quad = 0 \left. \right\} \quad \Longrightarrow \quad \begin{pmatrix} 1 & 1 \\ e^{kL} & e^{-kL} \end{pmatrix} \begin{pmatrix} c_1 \\ c_2 \end{pmatrix} = 0$$
$$x = L: c_1 e^{kL} + c_2 e^{-kL} = 0$$

$$\Longrightarrow \quad c_1 = c_2 = 0 \quad \Longrightarrow X(x) \equiv 0 \qquad \text{— the trivial solution again.}$$

**Case 3:** $\mu < 0$, that is $\mu = -k^2 < 0$.
Equation (3.9a) implies that $X'' + k^2 X = 0$, which is true if and only if $X(x) = c_1 \cos kx + c_2 \sin kx$. And Equation (3.10) gives us

$$X(0) = 0 \Longrightarrow \qquad c_1 . 1 = 0 \Longrightarrow c_1 = 0$$
$$X(L) = 0 \Longrightarrow c_2 \sin kL = 0 \Longrightarrow c_2 = 0 \text{ or } \sin kL = 0$$

A nontrivial solution is possible only if $kL = \pi, 2\pi, 3\pi, \ldots$, an integer multiple of $\pi$, and $c_2 \neq 0$.

The only nontrivial solutions are multiples of

$$X_n(x) = \sin \left( \frac{n\pi x}{L} \right), \quad n = 1, 2, 3, \ldots .$$

---

**Definition 3.16**
$$\mu = -k^2 = -\left( \frac{\pi}{L} \right)^2, -\left( \frac{2\pi}{L} \right)^2, -\left( \frac{3\pi}{L} \right)^2, \ldots \text{ are called } \textbf{eigenvalues}$$
*or* **characteristic values.**
$$X_n(x) = \sin \left( \frac{n\pi x}{L} \right), \quad n = 1, 2, 3, \ldots \quad \text{are called } \textbf{eigenfunctions}$$
*or* **characteristic functions.**

---

**Step 3:**

This choice of $\mu$ has implications for the solution of Equation (3.9b):

$$T''(t) + k^2 c^2 T(t) = 0, \quad k = \frac{\pi}{L}, \frac{2\pi}{L}, \frac{3\pi}{L}, \ldots$$

means that for each possible $k$ we get an independent solution. The coefficients of independent solutions will also depend on the values of $k$:

$$T_n(t) = A_n \cos \left( \frac{n\pi ct}{L} \right) + B_n \sin \left( \frac{n\pi ct}{L} \right), \quad n = 1, 2, 3, \ldots .$$

meaning that for each possible $k$ we get a solution of (3.6a) satisfying (3.6d).

$$u_n(x, t) = \left( A_n \cos \left( \frac{n\pi ct}{L} \right) + B_n \sin \left( \frac{n\pi ct}{L} \right) \right) \sin \left( \frac{n\pi x}{L} \right), \quad n = 1, 2, 3, \ldots .$$

Since all the infinite $n$ cases are possible solutions, called *harmonics*, the *principle of superposition* says that the most general solution is their linear combination. We have therefore an infinite series solution,

$$u = f(x,t) = \sum_{n=1}^{\infty} \sin\left(\frac{n\pi x}{L}\right)\left(A_n \cos\omega_n t + B_n \sin\omega_n t\right),$$

which describes the most general vibration of the ideal string fixed at its ends.

**Remarks**

* The boundary conditions determine the *allowed* $k$'s and $\omega$'s.

* $k_n = \dfrac{n\pi}{L} = \dfrac{2\pi}{\lambda_n}$,   where the $\lambda_n$ are the allowed *wavelengths*.

* $\omega_n = \dfrac{n\pi c}{L}$   are the allowed angular *frequencies*   — *eigenfrequencies*.

* $\omega_n = \dfrac{n\pi c}{L} = n\omega$. That is, $\omega_n$ is an integer multiple of the *fundamental frequency* $\omega = \dfrac{\pi c}{L}$   — the $n^{\text{th}}$ *harmonic overtone*.

* $u_n(x,t) = \left(A_n \cos\omega_n t + B_n \sin\omega_n t\right)\sin\left(\dfrac{n\pi x}{L}\right)$ describe the $n$th *normal mode* of vibration, the first few examples of which are shown in Figure 3.18.

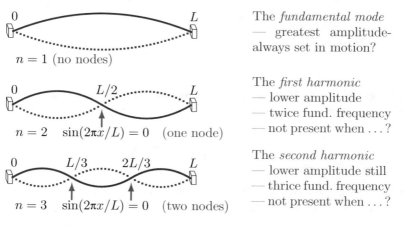

The *fundamental mode* — greatest amplitude- always set in motion?

The *first harmonic* — lower amplitude — twice fund. frequency — not present when ...?

The *second harmonic* — lower amplitude still — thrice fund. frequency — not present when ...?

$n = 1$ (no nodes)

$n = 2$   $\sin(2\pi x/L) = 0$   (one node)

$n = 3$   $\sin(2\pi x/L) = 0$   (two nodes)

**Figure 3.18**  The first few modes of vibration.

**Remarks** These are some points of a physical nature to note at this stage.

* If the string is pulled aside (plucked) anywhere along its length then the fundamental mode, shown at the top of Figure 3.18, will be present in the solution, since the fundamental mode features displacement at all points.

* On the other hand, a given harmonic is not present in the solution when the plucked point coincides with any of that harmonic's nodes. For example, if the string were to be plucked at its centre, the implied displacement would be inconsistent with that point being a node—a point of no motion. Consequently, the first harmonic mode (and any others that feature a node at the centre) cannot be included in the series expansion.

* $\dfrac{\omega_n}{k_n} = \dfrac{n\pi c}{L} \cdot \dfrac{L}{n\pi} = c = \sqrt{\dfrac{\tau}{\rho}}$       — the same speed for all modes.

**Step 4:**

To determine the unknown constants, $\{A_n\}_{n=1,2,\ldots,\infty}$ and $\{B_n\}_{n=1,2,\ldots,\infty}$, we apply the initial conditions of Equations (3.6b) and (3.6c). These imply that

$$f(x,0) = \sum_{n=1}^{\infty} A_n \sin\left(\frac{n\pi x}{L}\right) \equiv g(x), \qquad (3.11)$$

and

$$f_t(x,0) = \sum_{n=1}^{\infty} B_n.\omega_n \sin\left(\frac{n\pi x}{L}\right) = 0. \qquad (3.12)$$

Equation (3.12) tells us that

$$B_n = 0 \ \forall \ n,$$

and Equation (3.11) tells us that there exists a Fourier sine series for $g(x)$, with coefficients

$$A_n = \frac{2}{L} \int_0^L g(x) \sin\left(\frac{n\pi x}{L}\right) dx.$$

Given an initial form such as that represented in Figure 3.17, the $A_n$s can be determined and inserted in the series solution.

More general initial conditions of either plucked or struck strings are described by functions such as

$$\begin{cases} u(x,0) = u_0(x) \text{ — string shape at } t = 0, \\ u_t(x,0) = v_0(x) \text{ — string speed at } t = 0. \end{cases}$$

These conditions, or any linear combination of these conditions, are invoked to give $\{A_n, B_n\}$ in the more general case:

$$\begin{cases} A_n = \dfrac{2}{L} \displaystyle\int_0^L u_0(x) \sin\left(\dfrac{n\pi x}{L}\right) \mathrm{d}x \\ B_n = \dfrac{2}{\omega_n L} \displaystyle\int_0^L v_0(x) \sin\left(\dfrac{n\pi x}{L}\right) \mathrm{d}x \end{cases}$$

With all unknowns determined, the problem is solved.

**Remark**

* The string system just considered is a good model for a string attached to an electric guitar. The electric guitar body is (usually) solid. It is therefore appropriate to assume the string ends are fixed. The body itself undergoes very little vibration of its own (if any) while the string is vibrating. Consequently, the notes registered by an electric guitar are almost as pure as those determined mathematically.

  This differs fundamentally from the case of an acoustic guitar. The body of an acoustic guitar is hollow with the strings attached at one end to its flexible top plate. The vibrations of the strings are therefore transferred in part to the vibrations of the top plate (the ensuing air vibrations in the body are in fact responsible for a large proportion of the sound produced).

  To describe the vibrations of a string on an acoustic guitar it is therefore more reasonable to adopt the model of one end of the string attached to moveable mass which is subject to an elastic restoring force. However, even in this case it can be shown that discrete vibration modes arise. Although related by a mathematical formula, the frequencies of the higher-order modes are not simple multiples of a fundamental frequency. (See Supplementary problem 21.)

✎  **Mastery Check 3.12:**

For the problem of a string of length $L$ and density $\rho$ under tension $\tau$, fixed at its ends and pulled aside a distance $h$ at $x = b$, derive a closed-form expression for the string's energy of vibration

$$E = \frac{1}{2}\rho \int_0^L \left(\frac{\partial u}{\partial t}\right)^2 \mathrm{d}x + \frac{1}{2}\tau \int_0^L \left(\frac{\partial u}{\partial x}\right)^2 \mathrm{d}x,$$

$$\underbrace{\phantom{\frac{1}{2}\rho \int_0^L \left(\frac{\partial u}{\partial t}\right)^2 \mathrm{d}x}}_{\text{kinetic}} \quad \underbrace{\phantom{\frac{1}{2}\tau \int_0^L \left(\frac{\partial u}{\partial x}\right)^2 \mathrm{d}x}}_{\text{potential}}$$

using our separation of variables solution.

For the case $b = L/2$ show that the energy is conserved by comparing your result obtained above with the initial work done in pulling the string from its unstretched state.

✎

✎  **Mastery Check 3.13:**

Using the separation of variables method, derive the unique solution to the mixed b.v.p.

$$\begin{cases} \text{(a) } u_t(x,t) = k u_{xx}(x,t) & 0 < x < a, \ t > 0 \\ \text{(b) } u(0,t) = 0 & t \geq 0 \\ \text{(c) } u(a,t) = 0 & t \geq 0 \\ \text{(d) } u(x,0) = g(x) & 0 \leq x \leq a, \ \text{where } g(0) = g(a) = 0. \end{cases}$$

✎

# 3.F  Supplementary problems

**Section 3.A**

1. Use a Taylor series approximation to find the nature of any critical points for the function
$$f(x,y) = (x+2y)e^{-x^2-y^2},$$
for $\{(x,y) : x^2 + y^2 \le 1\}$.

2. Locate and classify the critical points of the following functions:

   (a) $f(x,y) = y^3 + 3x^2y - 3x^2 - 3y^2 + 2$,

   (b) $f(x,y) = x^3 - y^3 - 2xy + 6$,

   (c) $f(x,y) = e^{-x^2}\left(2xy + y^2\right)$.

**Section 3.B**

3. Determine the maximum and minimum values of
$$f(x,y) = x^2 + y^2 - 3x,$$
in the region $|x| + |y| \le 1$.

4. Show that in the region $|x| \le 1$, $|y| \le 1$ the function (Figure 3.19),
$$f(x,y) = e^{x^4 - y^4},$$
has a stationary point which is a saddle point, then determine its maximum and minimum values in the region.

5. Determine the extreme points of the surface
$$z = \sin x \sin y \sin(x+y),$$
over the rectangular domain
$$D = \{(x,y) : 0 \le x, y \le \pi\}.$$

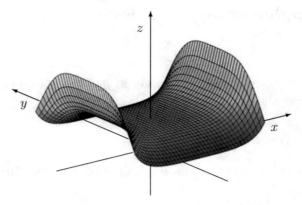

**Figure 3.19** The graph of $z = e^{x^4 - y^4}$.

6. Determine the maximum and minimum values of

$$f(x, y) = xy \ln(1 + x^2 + y^2),$$

in the region $x^2 + y^2 \leq 1$.

7. Determine the maximum and minimum values of

$$f(x, y) = (x + y)e^{-x^2 - y^2},$$

(Figure 3.20) in the region $x^2 + y^2 \leq 1$.

8. Determine the maximum and minimum values of

$$f(x, y) = (x + y)e^{-x^2 - y^2},$$

(Figure 3.20) in the triangular region:

$$R = \{(x, y) : x \geq 0, y \geq 0, x + y \leq 2\}.$$

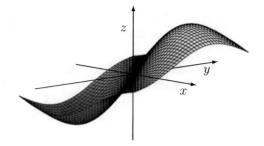

**Figure 3.20**   The graph of $z = (x + y)e^{-x^2 - y^2}$.

9. Determine the maximum of the function

$$f(x, y, z) = \log x + \log y + 3 \log z,$$

over that portion of the sphere $x^2 + y^2 + z^2 = 5r^2$, in the first octant. Deduce that

$$abc^3 \leq 27 \left( \frac{a + b + c}{5} \right)^5.$$

10. Suppose $C$ is the conic described by the equation

$$Ax^2 + 2Bxy + Cy^2 = 1,$$

where $A > 0$ and $B^2 < AC$. If we denote by $p$ and $P$ the distance from the origin to the closest and furthest point on the conic, respectively, show that

$$P^2 = \frac{A + C + \sqrt{(A - C)^2 + 4B^2}}{2(AC - B^2)},$$

with an analogous expression for $p^2$.

**Section** 3.C

11. Surveyors have measured two sides and the angle between them of a triangular plot of land, for the purpose of finding the area of the plot. The area is given by $f = \frac{1}{2}ab \sin C$, where $a$ and $b$ are the lengths of the sides and $C$ is the included angle. The measurements are all subject

to error. The measured values for the sides were $a = 152.60 \pm 0.005$ m and $b = 163.81 \pm 0.005$ m, and for the included angle $\theta = 43°26' \pm 0.2'$.

What is the largest possible error in the computation of the area? What is the largest possible percentage error?

## Section 3.D

12. Suppose we wish to represent the function $f(x) = x^2$, $-1 \le x \le 1$, by a combination of cosine functions of the form

$$g(x; a, b, c, d) = a + b\cos\pi x + c\cos 2\pi x + d\cos 3\pi x.$$

Find the least-squares values of $a, b, c, d$. Plot the two functions on the same graph to check the fit.

## Section 3.E

13. Show that $u(x, t) = \exp(-x^2/4t)/\sqrt{t}$ is a solution to the parabolic, 1D diffusion equation

$$\frac{\partial u}{\partial t} = \frac{\partial^2 u}{\partial x^2}.$$

14. Show that $u(x, y, t) = t^{-1}\exp(-(x^2 + y^2)/4t)$ is a solution to the parabolic, 2D diffusion equation

$$\frac{\partial u}{\partial t} = \frac{\partial^2 u}{\partial x^2} + \frac{\partial^2 u}{\partial y^2}.$$

15. Suppose $u : \mathbb{R}^3 \to \mathbb{R}$ is a $C^2$ function of variables $(x, y, z)$. By introducing 3D cylindrical polar coordinates (Section 1.D) confirm the expression for the Laplacian shown in the centre of Figure 3.13, applied to $U(r, \theta, z) = u(r\cos\theta, r\sin\theta, z)$.

16. Suppose $u : \mathbb{R}^3 \to \mathbb{R}$ is a $C^2$ function of variables $(x, y, z)$. By introducing 3D spherical polar coordinates (Section 1.D) confirm the expression for the Laplacian shown in the right-hand side of Figure 3.13, applied to $U(\rho, \phi, \theta) = u(\rho\sin\phi\cos\theta, \rho\sin\phi\sin\theta, \rho\cos\phi)$.

17. By introducing new variables $\xi = x + y$ and $\eta = \ln(x - y)$ (such that $x - y > 0$) and using the chain rule, determine all $C^2$ functions that solve the differential equation

$$(x - y)\left(\frac{\partial^2 u}{\partial x^2} + \frac{\partial^2 u}{\partial y^2}\right) = 4(x + y)^2.$$

18. Suppose $u : \mathbb{R}^2 \to \mathbb{R}$ is a $C^2$ function of variables $(x, t)$. With the help of the change of variables $\xi = x + ct$ and $\eta = x - ct$, transform the hyperbolic, 1D wave equation

$$\frac{\partial^2 u}{\partial x^2} - \frac{1}{c^2} \frac{\partial^2 u}{\partial t^2} = 0$$

into a simpler form involving a function $U(\xi, \eta)$ and thereby determine all possible $C^2$ functions that solve the wave equation.

19. By introducing the change of variables $\xi = x + e^y$ and $\eta = x - e^y$, determine the most general differentiable solution to the partial differential equation

$$e^{2y} \frac{\partial^2 u}{\partial x^2} - \frac{\partial^2 u}{\partial y^2} + \frac{\partial u}{\partial y} = 0.$$

20 Use the method of separation of variables to find the general solution to the equation

$$x^2 u_{xx} + x u_x - u_y = 0.$$

Find the particular solution that satisfies the boundary conditions $u(1, 0) = 0$, $u(e, 0) = 2$.

Hint: At some stage you will need to substitute $x = e^v$.

21 Redo the 1D wave (vibrating string) problem for the initial and boundary condition sets

(a) $(h_1, h_2, g_1, g_2) = (0, 0, 0, g(x))$, $0 \le x \le 2$;

(b) initial conditions $(g_1, g_2) = (g(x), 0)$, and boundary conditions

$$\left( f(0, t) = 0, M \frac{\partial^2 f}{\partial t^2}(L, t) = -\tau \frac{\partial f}{\partial x}(L, t) - \kappa f(L, t) \right) \text{ for } M, \kappa > 0.$$

Case (b) models a string of length $L$ under tension $\tau$ attached at $x = L$ to a mass $M$ subject to an elastic restoring force of strength $\kappa$.

# Chapter 4

# Integration of multivariable functions

We have seen that differentiation of a function is the result of a limit process. This led to the definition of a tangent line to a curve in 2D and a tangent plane to a surface in 3D. In this chapter we shall again consider limit processes but with the aim of establishing the reverse operation, that of integration of a scalar function of many variables.

It should be emphasized that we focus attention here on integrals of functions over subsets of $\mathbb{R}^n$. That is, the integrals we consider are taken over zero curvature regions in $\mathbb{R}^n$: a straight line segment in $\mathbb{R}$, a planar region in $\mathbb{R}^2$, a volume subset in $\mathbb{R}^3$, etc. In this important context we shall be able to rely on some familiar concepts derived from geometry such as areas (under curves) and volumes (under surfaces) to assist our understanding. In Chapter 5, we will revisit 1D and 2D integration but in the sense of integrals over nonzero curvature geometries (curves and curved surfaces in $\mathbb{R}^3$). In that context, geometric interpretations will be replaced with more physical conceptions.

## 4.A Multiple integrals

As we've done before, we shall first revisit the 1D case. To be precise, we shall brush up on the Riemann theory of integration for a function of one variable. Readers interested in the more general measure theory of Riemann-Stieltjes integration are referred to [2]. A comparison between the foundations of single-variable integration and multivariable integration is particularly fruitful, since the latter case is essentially a direct generalization of the former case.

© Springer Nature Switzerland AG 2020

S. J. Miklavcic, *An Illustrative Guide to Multivariable and Vector Calculus*,

https://doi.org/10.1007/978-3-030-33459-8_4

Almost every idea and theoretical result we discuss in this chapter is as valid for functions of an arbitrary number of variables as for functions of two variables. Therefore to simplify matters, we will present the theory (mostly) with reference to functions of *two* variables and discuss the more general case at the end of the chapter. Any important practical differences will be highlighted in that discussion.

As already mentioned, the definition of the integral of a function $f$ of $\boldsymbol{x} \in \mathbb{R}^n$ is based on a limit process. We illustrate the first steps of the process in Figure 4.1.

**Integration of $f : I \subset \mathbb{R} \longrightarrow \mathbb{R}$**

Suppose $f$ is a continuous function of $x$ and assume that the interval $I$ is closed and bounded and lying in the function domain, $D_f$. That is,

$$I = \{x : a \le x \le b\} \subset D_f.$$

The *graph* of $f$ is a curve in $\mathbb{R} \times \mathbb{R} = \mathbb{R}^2$ as shown in Figure 4.1 below.

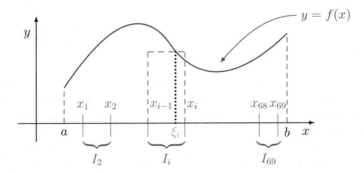

**Figure 4.1** The graph of $f$ and some subintervals of $I$.

First the interval $I$ is cut $I$ into small bits — this is called a *partition* of $I$:

$$a = x_0 < x_1 < x_2 < \ldots < x_{n-1} < x_n = b$$

with $n$ *subintervals* $I_i = \{x : x_{i-1} \le x \le x_i\}$, of width $\Delta x_i = x_i - x_{i-1}$. A few of these subintervals are shown in Figure 4.1.

Then, choosing some real number $\xi_i \in I_i$ from each subinterval, we form the sum

$$\sigma_n = \sum_{i=1}^{n} f(\xi_i)\Delta x_i.$$

This is called the *Riemann sum of $f$ over $I$*. From its construction we see that it *must* depend on the partition of $n$ subintervals.

**The geometric interpretation of $\sigma_n$ for $f : I \longrightarrow \mathbb{R}$.**

If $f \geq 0$, then $f(\xi_i)\Delta x_i$ is the *area* of the rectangle of height $f(\xi_i)$ and width $\Delta x_i$ as shown here in Figure 4.2.

Hence, the sum $\sigma_n$ is an approximation to the area "under" the curve $y = f(x)$ and over $I$.

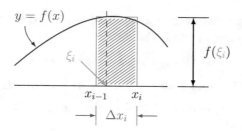

**Figure 4.2**  Rectangular area approximation.

To improve on this approximation we find numbers $\ell_i$ and $m_i$ in each interval $I_i$ such that $f(\ell_i) \leq f(x) \leq f(m_i)$ for all $x$ in $I_i$.

For a given partition we form the upper and lower sums

$$R_{\min} = \sum_{i=1}^{n} f(\ell_i)\Delta x_i \leq \sum_{i=1}^{n} f(\xi_i)\Delta x_i \leq \sum_{i=1}^{n} f(m_i)\Delta x_i = R_{\max}.$$

In this process we have constructed upper and lower bounds on $\sigma_n$. That is,

$$R_{\min} \leq \sigma_n \leq R_{\max}.$$

We now take the simultaneous limit of the number of intervals $n \to \infty$ and the representative size of the intervals $\max(\Delta x_i) \to 0$. We find that, as $n \to \infty$, $R_{\min}$ increases and $R_{\max}$ decreases. If the dual limits exist and $\lim R_{min} = \lim R_{max}$, then an application of a squeeze theorem gives:

---

**Definition 4.1**

*The integral of $f : \mathbb{R} \longrightarrow \mathbb{R}$ over $I$ is defined as the limit (if it exists)*

$$\lim_{\substack{n \to \infty \\ \max(\Delta x_i) \to 0}} \sigma_n = \int_a^b f(x)\,dx.$$

Under the conditions of continuity of $f$ and compactness of $I$ this limit does exist. This is a unique number, called the *Riemann integral* ([1], [2]), which is independent of how we set the partition originally.

Now let's see how things work for a function $f(x, y)$ of two variables.

**Integration of $f : R \subset \mathbb{R}^2 \longrightarrow \mathbb{R}$**

Suppose $f$ is a continuous function of $x$ and $y$ and, for starters, we are given the closed and bounded rectangular region

$$R = \{(x, y) : a \leq x \leq b, \ c \leq y \leq d\}$$

which lies inside $D_f$. Note that $R \subset D_f$ and $R$ is compact.

The *graph* of such a function $f$ is a *surface* in $\mathbb{R}^2 \times \mathbb{R} = \mathbb{R}^3$ as shown in Figure 4.3 below.

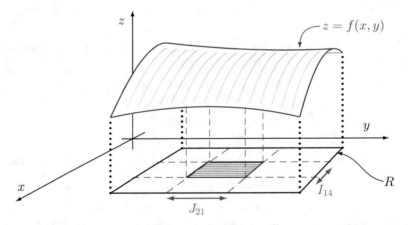

**Figure 4.3**  The graph of $f$ over rectangle $R$, and subrectangle $I_{14} \times J_{21}$.

The construction of the integral of $f$ over $R$ is accomplished in perfect harmony with the 1D case except that rectangles replace intervals.

First, the rectangle $R$ is partitioned into $n \times m$ rectangles $R_{ij} = I_i \times J_j$ of area $\Delta A_{ij}$, $1 \leq i \leq n, 1 \leq j \leq m$. We then choose some point $(\xi_i, \eta_j)$ from each rectangle $R_{ij}$ and form the sum

$$\sigma_{nm} = \sum_{i=1}^{n} \sum_{j=1}^{m} f(\xi_i, \eta_j) \Delta A_{ij}.$$

This is called the *Riemann sum of $f$ over $R$*. From its construction this Riemann sum *must* therefore depend on the partition.

**The geometric interpretation of $\sigma_{nm}$ for $f : R \subset \mathbb{R}^2 \longrightarrow \mathbb{R}$.**

If $f \geq 0$, then $f(\xi_i, \eta_j)\Delta A_{ij}$ is the *volume* of the rectangular block of height $f(\xi_i, \eta_j)$ and base area $\Delta A_{ij} = \Delta x_i.\Delta y_j$ as shown in Figure 4.4 below.

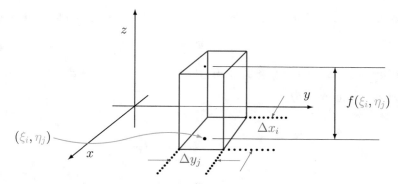

**Figure 4.4**   Rectangular block volume approximation.

Thus, the sum $\sigma_{nm}$ is an approximation to the volume "under" the surface $y = f(x, y)$ and over $R$.

A completely analogous line of reasoning to the steps leading up to the definition of a 1D integral can now be applied as follows.

For a given partition we form upper and lower sums of the rectangular blocks. Between these two sums is the sum $\sigma_{nm}$. With each refinement of the partition, new upper and lower sums are determined with the new lower sum increased compared with its predecessor and the new upper sum decreased compared with its predecessor.

If in the dual limit process of $n, m \to \infty$ and $\max\left(\sqrt{(\Delta x_i)^2 + (\Delta y_j)^2}\right) \longrightarrow 0$ the limits of upper and lower sums exist and are equal, an application of the squeeze theorem leads to our next definition.

---

**Definition 4.2**
*The **double integral** of $f : \mathbb{R}^2 \longrightarrow \mathbb{R}$ over $R$ is defined as the limit (if it exists)*

$$\lim_{\substack{m,n\to\infty \\ \max\left(\sqrt{(\Delta x_i)^2+(\Delta y_j)^2}\right)\to 0}} \sigma_{nm} = \iint_R f(x)\, \mathrm{d}A.$$

---

**An important theorem and some corollaries — 1D version.**

**Theorem 4.1** *All continuous functions are integrable on compact subsets of their domains of definition.*

**Corollary 4.1.1** *If $f \geq 0$, then $\int_I f(x)\,dx$ is the area under the curve $y = f(x)$.*

**Corollary 4.1.2** *If $f \geq g \geq 0$, then $\int_I \big(f(x) - g(x)\big)\,dx$ is the area between the curves (Figure 4.5).*

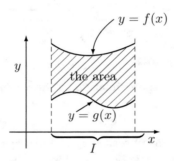

**Figure 4.5** The area between two curves in 2D.

**Corollary 4.1.3** *$\int_I 1\,dx$ is the length of interval $I$.*

**Corollary 4.1.4** *The average value of function $f$ over $I$ is*

$$\frac{1}{length\ I} \cdot \int_I f(x)\,dx \;=\; \frac{\displaystyle\int_I f(x)\,dx}{\displaystyle\int_I 1\,dx}.$$

**Corollary 4.1.5** *Linearity: If $a, b \in \mathbb{R}$ then*

$$\int_I \big(af(x) + bg(x)\big)\,dx \;=\; a\int_I f(x)\,dx + b\int_I g(x)\,dx.$$

**Corollary 4.1.6** *Additivity: (very important)*
*If $I \cap J = \{\} = \emptyset$, then $\displaystyle\int_{I \cup J} f(x)\,dx \;=\; \int_I f(x)\,dx + \int_J f(x)\,dx.$*

**An important theorem and some corollaries — 2D version.**

**Theorem 4.1** *All continuous functions are integrable on compact subsets of their domains of definition.*

**Corollary 4.1.1** *If $f \geq 0$, then $\iint_R f(x,y)\,\mathrm{d}A$ is the volume under the surface $z = f(x,y)$.*

**Corollary 4.1.2** *If $f \geq g \geq 0$, then $\iint_R \big(f(x,y) - g(x,y)\big)\,\mathrm{d}A$ is the volume of the body between the surfaces (Figure 4.6).*

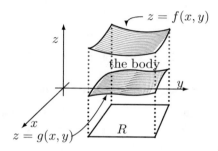

**Figure 4.6**  The volume between two surfaces in 3D.

**Corollary 4.1.3** $\iint_R 1\,\mathrm{d}A$ *is the area of $R$. — true not just for rectangles!!*

**Corollary 4.1.4**

$$\frac{1}{area\ R} \cdot \iint_R f(x,y)\,\mathrm{d}A \;=\; \frac{\iint_R f(x,y)\,\mathrm{d}A}{\iint_R 1\,\mathrm{d}A} \quad \text{is the average value of } f \text{ over } R.$$

**Corollary 4.1.5** *Linearity: If $a, b \in \mathbb{R}$ then*
$$\iint_R \big(af(x,y) + bg(x,y)\big)\,\mathrm{d}A \;=\; a\iint_R f(x,y)\,\mathrm{d}A + b\iint_R g(x,y)\,\mathrm{d}A.$$

**Corollary 4.1.6** *Additivity: (very important)*
If $R_1 \cap R_2 = \{\} \equiv \emptyset$, *then*
$$\iint_{R_1 \cup R_2} f(x,y)\,\mathrm{d}A \;=\; \iint_{R_1} f(x,y)\,\mathrm{d}A + \iint_{R_2} f(x,y)\,\mathrm{d}A.$$

Some comments on Theorem 4.1 are warranted.

From Theorem 1.2 all continuous functions on compact domains are bounded. These functions do not exhibit singular behaviour and so in the integral definition all partitions have finite volume (area). The unique limits in Definitions 4.1 and 4.2 therefore exist.

Corollary 4.1.3 may seem trivial but is undeniably useful in some contexts. (See Mastery Check 5.24, Section 5.D, and Example 5.9.)

Corollary 4.1.5 allows us to split complex functions into sums of simpler functions and to integrate them individually.

Corollary 4.1.6 is useful when an integration domain is or can be described piecewise, especially if the pieces warrant different techniques of integration.

# 4.B   Iterated integration in $\mathbb{R}^2$

Iterated integration is the workhorse of multiple integrals.

The definition of the multiple integral as the limit of a sum is not practical. Fortunately, there is an alternative. The suggestion is made that we calculate our "volumes" by *slicing* rather than by *dicing*.

Consider the thin slice of the "body" under $f$ shown in Figure 4.7. The area of the left-hand side face, that is, the area under the curve of constant $y$, $y = y_0$, is $A(y_0) = \int_a^b f(x, y_0)\, \mathrm{d}x$. Similarly, $A(y_0 + \Delta y) = \int_a^b f(x, y_0 + \Delta y)\, \mathrm{d}x$ is the area of the right-hand side face.

If $|\Delta y|$ is a small increment then $A(y_0) \approx A(y_0 + \Delta y)$, which is easy to see by expanding $f(x, y_0 + \Delta y)$ in a Taylor series about $(x, y_0)$. Then, using the simple two-point trapezoidal rule approximation, the volume of the "slice" is approximately

$$V(y_0) = \frac{1}{2}\big(A(y_0) + A(y_0 + \Delta y)\big)\Delta y$$
$$= A(y_0)\Delta y + O(\Delta y^2).$$

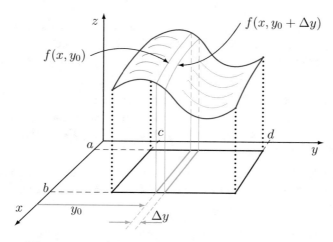

**Figure 4.7**  Adding slices to determine volumes.

The total volume of the "body" under $f(x, y)$ is then the limiting sum of these 1D volumes of slices (Definition 4.1) as $\Delta y \to 0$. That is, the volume under $f(x, y)$ over $R$ is the Riemann integral of $A(y)$ over the interval $c \le y \le d$:

$$V = \int_c^d A(y) \, dy = \int_c^d \left( \int_a^b f(x, y) \, dx \right) dy.$$

Alternatively, slicing parallel to the $y$-axis instead of the above would give

$$V = \int_a^b A(x) \, dx = \int_a^b \left( \int_c^d f(x, y) \, dy \right) dx,$$

which must give the exact same value for the volume.

Hence, for integration over the rectangle $[a, b] \times [c, d]$ we have the important result

$$\underbrace{\iint_R f(x, y) \, dA}_{\substack{\text{double integral of} \\ f \text{ over } R}} = \underbrace{\int_a^b \left( \int_c^d f(x, y) \, dy \right) dx = \int_c^d \left( \int_a^b f(x, y) \, dx \right) dy}_{\text{iterated integrals of } f \text{ over } R}$$

The left-hand side is the definition of a double integral, while the two right-hand sides are the actual ways one can evaluate the double integral. These are called the *iterated* integrals of $f$. In each case of iterated integral, the *inner* integral within the parentheses is to be evaluated first, only then is the *outer* integral evaluated. Notice how in the above equation the integration limits follow the variable being integrated.

At a very practical level, when evaluating the inner integral we treat the *outer integral variable* as if it were a constant parameter. In such a situation all single-variable techniques apply to each individual iterated integral. Example 4.1 illustrates this process.

Regarding notation, the above clearly specifies the order of operation. However, to skip the parentheses (and avoid the tedium of writing them) we have two alternative notations in common use:

$$\int_a^b \int_c^d f(x, y) \, dy \, dx$$    — this borders on the ambiguous; the user must not confuse the order.

$$\int_a^b dx \int_c^d f(x, y) \, dy$$    — this is somewhat better; it is easier to interpret and better for complex regions (see next section).

## ■  Example 4.1:

Determine the volume of the body lying under the surface $z = x^2 + y^2$ (Figure 4.8) and over the rectangle,

$$R = \{(x, y) : 0 \leq x \leq 1, \ 0 \leq y \leq 1\}, \ ( = [0, 1] \times [0, 1]).$$

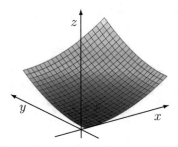

**Figure 4.8**  The graph of $z = x^2 + y^2$.

**Solution:**

$$V = \iint_R (x^2 + y^2)\,\mathrm{d}A \qquad \text{— the double integral}$$

$$= \int_0^1 \mathrm{d}y \int_0^1 (x^2 + y^2)\,\mathrm{d}x \qquad \text{— an iterated integral version}$$

$$= \int_0^1 \mathrm{d}y \left[\frac{x^3}{3} + y^2 x\right]_{x=0}^1 \qquad \text{—} \, y \text{ is held "fixed" during the } x\text{-integration}$$

$$= \int_0^1 \left(\frac{1}{3} + y^2\right)\mathrm{d}y = \left[\frac{y}{3} + \frac{y^3}{3}\right]_0^1 = \frac{2}{3} \text{ volume units.}$$

In this iterated integral $y$ is given the role of the outer integration variable, while $x$ plays the role of inner variable. We could just as easily have reversed the roles to arrive at the same result. Notice the very important fact, which will be contrasted with later, that in this example the bounds on the $x$-integral are constants; they are *not* functions of $y$!

■

✎  **Mastery Check 4.1:**

Evaluate $\displaystyle\iint_R y\mathrm{e}^{xy}\,\mathrm{d}A$, where $R = \{(x,y) : 0 \le x \le 1, 0 \le y \le 2)\}$.

✎

✎  **Mastery Check 4.2:**

Compute $\displaystyle\iint_R \frac{x\mathrm{e}^{x\sqrt{y}}}{\sqrt{y}}\,\mathrm{d}A$, where $R = \{(x,y) : 0 < x < 1, 0 < y < 1)\}$.

✎

# 4.C   Integration over complex domains

In general practice a region of integration is more often *not* a rectangle. We therefore need a way to work with compact regions that are more complex. Luckily, we can treat this problem using our previous results, but after first rethinking the function, the domain, and the iteration definitions.

First, let's look at the function and its domain. For a function $f$ defined on a non-rectangular, compact domain $D$, we introduce a new function and new domain through the following definition.

**Definition 4.3**

*Suppose $f(x, y)$ is continuous over a compact domain $D \subseteq D_f$. Let $\hat{f}$ be the **extension** of $f$ beyond $D$:*

$$\hat{f}(x, y) = \begin{cases} f(x, y) \text{ for } (x, y) \in D \\ 0 \text{ for } (x, y) \notin D. \end{cases}$$

Since $D$ is bounded there exists a rectangle $R$ such that $D \subset R$. Thus, if $\hat{f}$ is integrable over $R$, then

$$\underbrace{\iint_D f(x, y)\, \mathrm{d}A}_{\text{definition}} = \underbrace{\iint_R \hat{f}(x, y)\, \mathrm{d}A}_{\text{calculable value}}$$

This last equation is true since all we have done is added zero to the original double integral. The picture we imagine is something like Figure 4.9 below.

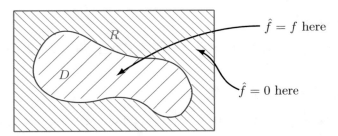

**Figure 4.9** The extended function $\hat{f}$ and its domain $R$.

Second, we examine the domain, $D$, a little further. What can these more complicated domains look like? In Figures 4.10–4.12 we define three main classes of regions into one class of which the domain $D$, or a piece of $D$, may be placed.

## Type I:

Suppose the domain $D$ is of the kind shown in Figure 4.10 and defined as

$$D = \{(x, y) : a \leq x \leq b, g_1(x) \leq y \leq g_2(x)\}.$$

This is called a *y-simple domain*, with the variable $y$ bounded by two functions of $x$.

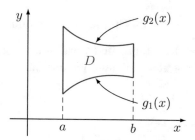

**Figure 4.10**   A $y$-simple domain.

## Type II:

Suppose the domain $D$ is of the kind shown in Figure 4.11 and defined as

$$D = \{(x, y) : c \leq y \leq d, h_1(y) \leq x \leq h_2(y)\}.$$

This is called a *x-simple domain*, with the variable $x$ bounded by two functions of $y$.

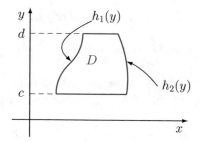

**Figure 4.11**   An $x$-simple domain.

## Type III:

This is a domain $D$ with an appearance like that shown in Figure 4.12, and with enough flexibility to have two interpretations. It could either be treated as a special case of **I** with $g_1(a) = g_2(a)$ and $g_1(b) = g_2(b)$, or as a special case of **II** with $h_1(c) = h_2(c)$ and $h_1(d) = h_2(d)$. That is, this sort of domain could be either $x$-simple or $y$-simple depending on how it is described.

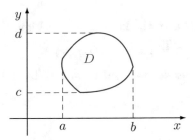

**Figure 4.12**  A domain that is both $x$-simple and $y$-simple.

Third, we bring these ideas together to arrive at a strategy for evaluating integrals over non-rectangular domains. We demonstrate this with a *y-simple domain* (Fig. 4.13).

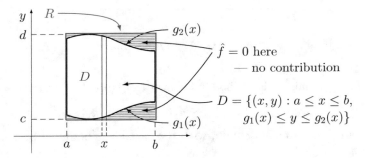

**Figure 4.13**  A $y$-simple domain in a rectangle $R$.

By construction we have the iterated integral of $\hat{f}$ over $R = [a, b] \times [c, d] \supset D$,

$$\iint_D f(x, y)\, \mathrm{d}A = \iint_R \hat{f}(x, y)\, \mathrm{d}A = \int_a^b \mathrm{d}x \int_c^d \hat{f}(x, y)\, \mathrm{d}y.$$

However, for every value of the outer integral variable $x$, $\hat{f} = 0$ outside the interval $g_1(x) \le y \le g_2(x)$, and $\hat{f} = f$ in the interior of that interval. Hence,

$$\iint_D f(x, y)\, \mathrm{d}A = \underbrace{\int_a^b \mathrm{d}x \int_{g_1(x)}^{g_2(x)} f(x, y)\, \mathrm{d}y}.$$

iterated integral of $f$ over $D$

We can now invoke the contrast alluded to earlier regarding the variable dependence of the limits of the inner integral. For all cases of non-rectangular domains, the limits of the inner integral will be functions of the outer integral variable. In the above example, the limits on the inner integral depend on $x$

and do *not* depend on $y$. Only in the case of rectangular domains will the limits of both the inner and outer integrals be constants!

For an *x-simple domain*, with $D = \{(x, y) : c \leq y \leq d, \ h_1(y) \leq x \leq h_2(y)\}$, we get the analogous result

$$\iint_D f(x, y) \, \mathrm{d}A = \int_c^d \mathrm{d}y \int_{h_1(y)}^{h_2(y)} f(x, y) \, \mathrm{d}x.$$

Once again, the limits of the inner integral depend on the outer integral variable; the limits of the inner integral here depend on $y$ and do *not* depend on $x$.

Through a very simple development we have arrived at very natural generalizations of iterated integrals over rectangles. Moreover, in the process we have done away with the extensions we used in this development.

The reader should now bear two things in mind. First, the order in which the iterated integrals are to be performed must be strictly adhered to. Second, interchanging the order will *always* involve a change in the limits. This is illustrated in Example 4.2 wherein a double integral is evaluated in two ways. The reader should note the limits on the two inner integrals and how they come about (see the vertical and horizontal bars in Figure 4.14).

## ■ Example 4.2:

Suppose $D$ is that region bounded by the lines $y = x$, $y = 2x$, and $x = 1$.

Calculate the area of $D$ as $\displaystyle\iint_D 1 \, \mathrm{d}A$.

It cannot be stressed enough that one should always start solving a multiple integral problem by drawing a picture of the *region* in question. Here it is the one shown in Figure 4.14. Sketching the region of integration, if it is possible, not only allows us to get some perspective on the task involved, it also allows us to determine the limits of the integration variables. More importantly, it can potentially identify difficulties with setting limits, which formulae themselves may not do. An example of such complication arises when we treat $D$ as $x$-simple.

The domain is both $y$-simple and piecewise $x$-simple.

We first treat $D$ as $y$-simple.

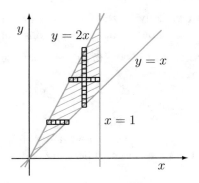

**Figure 4.14** The $y$-simple domain $D$.

$$\text{Area of } D = \iint_D 1 \, dA \;=\; \int_0^1 dx \int_x^{2x} 1 \, dy$$

$$= \int_0^1 dx \, \big[y\big]_x^{2x} \;=\; \int_0^1 x \, dx \;=\; \left[\frac{x^2}{2}\right]_0^1 = \frac{1}{2} \text{ area units.}$$

Now we treat $D$ as $x$-simple. This is slightly more complex as Figure 4.14 suggests. We must break $D$ up into two non-overlapping domains and invoke Corollary 4.1.6.

$$\text{Area of } D = \iint_D 1 \, dA \;=\; \int_0^1 dy \int_{y/2}^y 1 \, dx + \int_1^2 dy \int_{y/2}^1 1 \, dx$$

$$= \int_0^1 dy \, \big[x\big]_{y/2}^y + \int_1^2 dy \, \big[x\big]_{y/2}^1$$

$$= \int_0^1 \frac{y}{2} \, dy + \int_1^2 \left(1 - \frac{y}{2}\right) dy \;=\; \left[\frac{y^2}{4}\right]_0^1 + \left[y - \frac{y^2}{4}\right]_1^2$$

$$= \frac{1}{4} + (2 - 1) - \left(1 - \frac{1}{4}\right) \;=\; \frac{1}{2} \text{ area units.} \qquad\blacksquare$$

✍ **Mastery Check 4.3:**

Determine $\iint_D \left(xy + y^2\right) dA$, over each of the following domains $D$.

(a) $D = \{(x, y) : 0 \le x \le 1, \; 0 \le y \le x^2\}$.

(b) $D = \{(x,y) : 0 \leq x \leq \sqrt{y},\ 0 \leq y \leq 1\}.$

✎

✎ **Mastery Check 4.4:**

Evaluate the iterated integral $I = \int_0^1 dx \int_{\sqrt{x}}^1 e^{y^3}\, dy.$

Hint: If you have trouble with this iterated integral, try thinking about it first as a double integral over some domain $D$.

✎

# 4.D   Generalized (improper) integrals in $\mathbb{R}^2$

The iterated integral approach only works if the corresponding multiple integral exists (that is, the limit of the sum $\sigma_{mn}$ exists). It is very important to remember that thus far we have relied on the convenient assumptions of the function being continuous and the domain being bounded. However, in other cases the Riemann theory of integration on which the double integral was founded can break down. This leads us to consider so-called *improper integrals*, of which there are two types:

(a) One type involves *unbounded domains*, e.g. $D = \{(x,y) : x > 0,\ y > 0\}.$

(b) One type involves functions $f$, *not defined on part of the boundary*. For example, $f(x,y) = 1/\left(x^2 + y^2\right)$ is not defined at $(0,0)$.

Under these more general circumstances, to answer the question

$$\text{``Does } \iint_D f(x,y)\, dA \text{ exist?''},$$

we can rely somewhat on the combined action of the following theorems.

---

**Theorem 4.2**

*If* ...                                                              *then* ...

$$\left.\begin{array}{c} \int dx \int |f(x,y)|dy \\[2mm] and \int dy \int |f(x,y)|dx \end{array}\right\} \begin{array}{l} exist \\ and\ are \\ equal \end{array} \qquad \iint_D |f(x,y)|\, dA\ \ exists.$$

**Theorem 4.3**

$$\text{If} \quad \iint_D |f(x,y)|\,\mathrm{d}A \quad \text{exists, then} \quad \iint_D f(x,y)\,\mathrm{d}A \quad \text{exists.}$$

These two theorems may seem universally useful, and they are when any two iterated integrals *do not* give the same result, or if an iterated integral fails to converge. We would know then that the multiple integral does not exist. However, if only one ordered interated integral can be evaluated, and it converges, and we cannot evaluate the other iterated integral, then there may still remain some doubt about the existence of the multiple integral.

**Improper integrals in analysis**

Nevertheless, guided by the above theorems we jump straight to the task of evaluating the iterated integral forms of a generalized multiple integral. In tackling improper iterated integrals we take advantage of the wisdom of single-variable calculus.

In single-variable calculus, the improper definite integral $\int_I f(x)\,\mathrm{d}x$, for $f(x) > 0$ over the domain $I$, exists if

$$\lim_{\epsilon \to 0^+} J(\epsilon) = \lim_{\epsilon \to 0^+} \int_{a+\epsilon}^{b} f(x)\,\mathrm{d}x, \ a < b \ \textit{exists},$$ when $f$ is a function which diverges at the integration limit $x = a \in \overline{I}$ (Figure 4.15(a)); or

$$\lim_{R \to \infty} J(R) = \lim_{R \to \infty} \int_{a}^{R} f(x)\,\mathrm{d}x$$ exists, when $I$ is the unbounded domain, $I = [a, \infty)$ (Figure 4.15(b)).

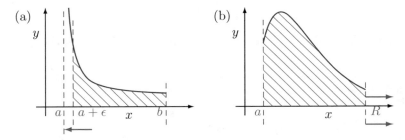

**Figure 4.15** Two types of improper integral in single-variable calculus.

In Figure 4.15(a) the function $f$ is singular at the lower limit $x = a$, while in Figure 4.15(b) the domain $I$ is unbounded.

We notice that both cases rely on the Riemann theory of integration of a *continuous* function defined on a *bounded* sub-domain to give a finite number $J(\epsilon)$ and $J(R)$, respectively. And both cases subsequently test the convergence of limits, the first $\lim_{\epsilon \to 0} J(\epsilon)$ and the second $\lim_{R \to \infty} J(R)$, respectively, to define and provide the integrals wanted.

In multivariable calculus the improper multiple integral

$$\iint_D f(x,y)\, dA \quad (\text{for } f(x,y) > 0)$$

is similarly to be identified as a suitable limit. In practice we work with the iterated integral version of this double integral. However, the multiple integral version of the principle is more easily described using the double integral itself.

The student reader will no doubt note the similarity between the two scenarios. The common denominator is the integration domain $D$ and in each case the sequence of smaller domains that do not present any difficulty. Note that the arguments below are valid and can substantiated *only* for the case of functions $f(x,y)$ that do not change sign over the integration region.

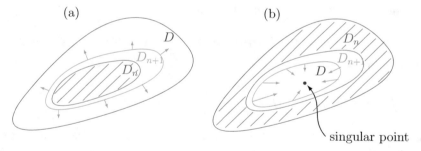

**Figure 4.16** (a) Finite domains converging to unbounded $D$;
(b) Regular domains converging to singular $D$.

Suppose $D$ is an *unbounded* domain, but contains *bounded* subsets, $D_n$, represented in Figure 4.16(a) such that

$$D_n \subset D_{n+1} \subset D \; \forall n \text{ and } \bigcup_{n=1}^{\infty} D_n = D.$$

Every point in $D$ belongs to at least one of the $D_n$.

On the other hand, suppose $D$ is a domain containing a *singular point* of $f$, but which contains *bounded* subsets, $D_n$, as illustrated in Figure 4.16(b), that *exclude* that point.

$$D_n \subset D_{n+1} \subset D \; \forall n \text{ and } \bigcup_{n=1}^{\infty} D_n = D.$$

Every point in $D$ belongs to at least one of the $D_n$.

In either case we have the following useful result.

---

**Theorem 4.4**

Let $f$ be a non-negative or non-positive function defined on a domain, $D$, which may be unbounded or contain points on which $f$ is undefined. For the sequence of bounded subsets, $D_n$, satisfying the conditions $D_n \subset D_{n+1} \subset D$ and $\bigcup_{n=1}^{\infty} D_n = D$, if

$$\lim_{n \to \infty} J(n) = \lim_{n \to \infty} \iint_{D_n} f(x,y) \, dA$$

exists, then the improper integral $\iint_D f(x,y) dA$ exists and is equal to this limit.

---

An immediate corollary of this is the following.

**Corollary 4.4.1** *For the function and domain conditions of Theorem 4.4, if the improper integral* $\iint_D f(x,y) \, dA$ *exists, then the iterated integrals*

$$\int dx \int f(x,y) \, dy \quad and \quad \int dy \int f(x,y) \, dx \quad \text{exist and are equal.}$$

Unfortunately, the only assertion that can be made for functions that change sign over the integral domain is Theorem 4.3. (See Mastery Check 4.8.) As we said, in practice we work with iterated integrals to determine our $J(n)$ integrals, then we take the limits of these, as illustrated by the next example.

■ **Example 4.3:**
Check whether the following integrals converge. If they do, compute them.

(a) $\displaystyle\iint_D \frac{dx \, dy}{1 + (x+y)^2}$, $D = \{(x,y) : x > 0, y > 0\}$.

(b) $\displaystyle\iint_D \frac{dx \, dy}{\sqrt{xy}}$, $D = \{(x,y) : 0 < x < 1, 0 < y < 1\}$.

**Solution:**

(a) Take the $x$-simple integral over a finite domain
$D_{AB} = \{(x, y) : 0 < x < A, 0 < y < B\}$.

$$\iint_{D_{AB}} \frac{1}{1 + (x + y)^2}\, dx\, dy = \int_0^B dy \int_0^A \frac{1}{1 + (x + y)^2}\, dx$$

$$= \int_0^B dy \Big[ \arctan(x + y) \Big]_{x=0}^A = \int_0^B \big( \arctan(y + A) - \arctan y \big)\, dy.$$

Now $\displaystyle \lim_{B \to \infty} \int_0^B \arctan y\, dy = \lim_{B \to \infty} \Big[ y \arctan y - \frac{1}{2} \ln(1 + y^2) \Big]_0^B$,

which evidently does not exist. The integral over $D$ does not converge.

(b) The integrand is undefined on the boundary of $D$, at $x = 0$ and at $y=0$.
So integrate over the domain $D_\epsilon = \{(x, y) : \epsilon < x < 1, \epsilon < y < 1\}$.

$$\iint_{D_\epsilon} \frac{1}{\sqrt{xy}}\, dx\, dy = \int_\epsilon^1 dy \int_\epsilon^1 \frac{1}{\sqrt{xy}}\, dx = \int_\epsilon^1 dy \Big[ 2\sqrt{\frac{x}{y}} \Big]_\epsilon^1$$

$$= \int_\epsilon^1 2\frac{1 - \sqrt{\epsilon}}{\sqrt{y}}\, dy = 4(1 - \sqrt{\epsilon})^2 \longrightarrow 4 \text{ as } \epsilon \to 0.$$

∎

## ✍ Mastery Check 4.5:

Does the double integral $I = \displaystyle\iint_D f(x, y)\, dA$ converge or diverge when

$$f(x, y) = \frac{1}{x^2 + y^2} \quad \text{and} \quad D = \{x, y) : x \geq 1,\ 0 \leq y \leq x\}?$$

Hint: Draw a picture. Consider the sub-domain $D_R = \{x, y) : 1 \leq x \leq R,$
$0 \leq y \leq x\}$, and let $R \to \infty$. Write out both iterated integral versions, and
choose the simpler of the two for analysis.

✍

## ✍ Mastery Check 4.6:
Dangers in $\infty$!

Consider the two iterated integrals

$$I_1 = \int_0^1 dy \int_1^\infty \big( e^{-xy} - 2e^{-2xy} \big)\, dx, \quad \text{and} \quad I_2 = \int_1^\infty dx \int_0^1 \big( e^{-xy} - 2e^{-2xy} \big)\, dy.$$

Show that one of these must be $< 0$, while the other must be $> 0$, and ponder
the implications in the context of Theorem 4.3.

✍

# 4.E  Change of variables in $\mathbb{R}^2$

Sometimes, the multiple integrals are not so easy to evaluate when expressed in Cartesian coordinates, $x$ and $y$. The problem might stem from the function being difficult, or from the domain being convoluted, or *both*. This is different from the 1D case where we only need to worry about the function.

In the 2D situation the problem could be avoided by changing variables from $x$ and $y$ to new variables, $u$ and $v$, say. Here, we discuss a process for doing this, and on the way we point out when it is possible to do so, and when it is not. The 1D case provides an interesting comparison.

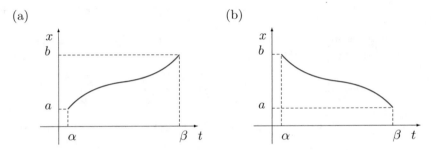

**Figure 4.17**  (a) $x(t)$ increasing; (b) $x(t)$ decreasing.

**Change of variables in single integrals**

**Aim:** We want to evaluate $\displaystyle\int_a^b f(x)\,\mathrm{d}x$ by invoking a transform $x = x(t)$.

Suppose $x(t)$ is strictly *increasing* as in Figure 4.17(a), then $x'(t) > 0$, and $\dfrac{\mathrm{d}F}{\mathrm{d}x} = f$. Then

$$\int_\alpha^\beta f(x(t))x'(t)\,\mathrm{d}t = \int_\alpha^\beta F'(x(t))x'(t)\,\mathrm{d}t$$

$$= \int_\alpha^\beta \frac{\mathrm{d}}{\mathrm{d}t}F(x(t))\,\mathrm{d}t = F(x(\beta)) - F(x(\alpha))$$

$$= F(b) - F(a) = \int_a^b f(x)\,\mathrm{d}x.$$

On the other hand suppose $x(t)$ is strictly *decreasing* as in the case shown in

Figure 4.17(b), then $x'(t) < 0$, and $\dfrac{\mathrm{d}F}{\mathrm{d}x} = f$. Then

$$\int_\alpha^\beta f(x(t))\big(-x'(t)\big)\,\mathrm{d}t = -\int_\alpha^\beta F'(x(t))x'(t)\,\mathrm{d}t$$

$$= \int_\beta^\alpha \frac{\mathrm{d}}{\mathrm{d}t}F(x(t))\,\mathrm{d}t = F(x(\alpha)) - F(x(\beta))$$

$$= F(b) - F(a) = \int_a^b f(x)\,\mathrm{d}x.$$

Consequently, $$\int_a^b f(x)\,\mathrm{d}x = \int_\alpha^\beta f(x(t))\left|\frac{\mathrm{d}x}{\mathrm{d}t}\right|\,\mathrm{d}t.$$

So, in this case we see the integration interval has changed, but more significantly the change of variable $x \longrightarrow t$ has introduced a positive factor $|x'|$ in the integral. This is called the *scale factor* since it scales up or down the interval size. We should expect a similar factor to appear in multiple integrals.

### Change of variables in double integrals

For convenience we shall consider only bijective transformations:

$$\tau : \boldsymbol{u} \mapsto \boldsymbol{x}(\boldsymbol{u}) = \begin{cases} x = x(u,v) \\ y = y(u,v) \end{cases}$$

such that $\dfrac{\partial(x,y)}{\partial(u,v)} \neq 0$. The Jacobian determinant (Definition 2.9) is involved in the transformation of double integrals:

$$\iint_D f(x,y)\mathrm{d}A \qquad \text{becomes expressed as} \qquad \iint_E g(u,v)\mathrm{d}A'.$$

Geometrically the transformation affects areas both globally and locally.

To see how the transformation does this consider in Figure 4.18 the "parallelogram" in the $xy$-plane created by constant $u$ and $v$ contours. Suppose opposite sides of the parallelogram are separated by differences $\mathrm{d}u$ and $\mathrm{d}v$.

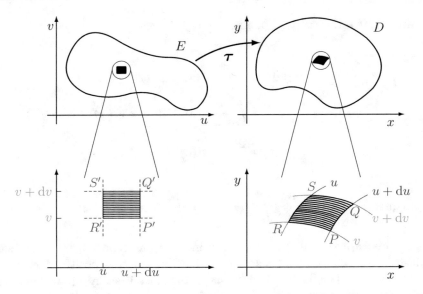

**Figure 4.18** A geometrical view of a change of variables.

For small $du$ and $dv$, the area of the element in $D$ is given by the vector product

$$dA = \left| \overrightarrow{RP} \times \overrightarrow{RS} \right| \qquad \text{— geometric interpretation (see Page 3)}$$

$$= \left| \underbrace{(dx_u\, e_1 + dy_u\, e_2)}_{\substack{\text{along a line of} \\ \text{constant } v}} \times \underbrace{(dx_v\, e_1 + dy_v\, e_2)}_{\substack{\text{along a line of} \\ \text{constant } u}} \right|.$$

Thus,

$$dA = \left| \left( \frac{\partial x}{\partial u}\, du\, e_1 + \frac{\partial y}{\partial u}\, du\, e_2 \right) \times \left( \frac{\partial x}{\partial v}\, dv\, e_1 + \frac{\partial y}{\partial v}\, dv\, e_2 \right) \right|$$
$$\text{— by the chain rule (Section 2.G)}$$

$$= \left| \frac{\partial(x, y)}{\partial(u, v)} \right| du\, dv \quad \text{— cross product gives the Jacobian determinant}$$

The reader should make particular note that it is the *absolute value* of the Jacobian determinant that appears here. This is reasonable since what we have done is transformed one area element to another, preserving the sign.

And so we have the following important theorem.

---

**Theorem 4.5**

Let $x(u,v)$ and $y(u,v)$ be a bijective [**one-to-one** and **onto**] and $C^1$ transformation of $E$ in the $uv$-plane onto $D$ in the $xy$-plane.

If $f(x,y)$ is integrable in $D$ then $f(x(u,v), y(u,v)) = F(u,v)$ is integrable in $E$ and

$$\iint_D f(x,y)\,\mathrm{d}A = \iint_E F(u,v)\left|\frac{\partial(x,y)}{\partial(u,v)}\right|\mathrm{d}A'.$$

---

**Remarks**

* The *absolute value* of the Jacobian is the *scale factor* between the two area elements, $\mathrm{d}A$ in the $xy$-plane and $\mathrm{d}A'$ in the $uv$-plane; it takes the role played by $\left|\dfrac{\mathrm{d}x}{\mathrm{d}t}\right|$ in the single-variable case.

  In other words
  $$\mathrm{d}A \;=\; \left|\frac{\partial(x,y)}{\partial(u,v)}\right|\mathrm{d}A'.$$
  $\underbrace{\phantom{xxxxxxxx}}$
  an area element in $(x,y)$ $\qquad$ $\underbrace{\phantom{xxxxxxxx}}$ an area element in $(u,v)$

* If $x(u,v)$ and $y(u,v)$ are bijective transformations, then $\dfrac{\partial(x,y)}{\partial(u,v)} \neq 0$.

* If $x(u,v)$ and $y(u,v)$ are $C^1$, then $F(u,v) = f(x(u,v), y(u,v))$ is integrable over $E$ whenever $f$ is integrable over $D$.

* A change of variables in a *double* integral is NOT the same as a substitution in an *iterated* integral.

Before we demonstrate how one considers the change of variable to evaluate a double integral, we encourage the reader to verify the following Jacobian expressions.

## ✍ Mastery Check 4.7:

Show that in transforming from (Hint: See Section 1.D.)

(a) Cartesian to polar coordinates $(r, \theta)$ the Jacobian is $r$;

(b) Cartesian to cylindrical coordinates $(r, \theta, z)$ the Jacobian is $r$;

(c) Cartesian to spherical coordinates $(\rho, \phi, \theta)$ the Jacobian is $\rho^2 \sin \phi$.

## ■ Example 4.4:

We wish to show that the volume of the right cone of Figure 4.19, of radius $a$ and height $h$, is $V = \dfrac{1}{3}\pi a^2 h$.

We do this by integrating a variable height function $z = \dfrac{h}{a}\left(a - \sqrt{x^2 + y^2}\right)$ over the region $R : x^2 + y^2 \leq a^2$.

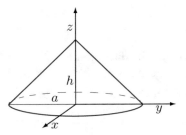

**Figure 4.19** A right cone.

The volume is $V = \displaystyle\iint_R \dfrac{h}{a}\left(a - \sqrt{x^2 + y^2}\right) \mathrm{d}x\,\mathrm{d}y$.

Change to polar coordinates $(x, y) \longrightarrow (r, \theta)$, $x = r\cos\theta$, $y = r\sin\theta$, $J = r$.

$$V = \iint_R \frac{h}{a}(a - r)r\,\mathrm{d}r\,\mathrm{d}\theta = \int_0^a \mathrm{d}r \int_0^{2\pi} \frac{h}{a}(ar - r^2)\,\mathrm{d}\theta$$

$$= \frac{2\pi h}{a}\left[a\frac{r^2}{2} - \frac{r^3}{3}\right]_0^a = \frac{\pi h a^2}{3}.$$

■

## ✍ Mastery Check 4.8:

Does the integral $\displaystyle\iint_{\mathbb{R}^2} \dfrac{\mathrm{d}x\,\mathrm{d}y}{(1 + x^2 + y^2)^2}$ converge? If it does, find its value.

Hint: Find the integral over $D = \{(x, y) : x^2 + y^2 < A^2\}$, and let $A \to \infty$.

✍

## Philosophy of a change of variables

It is worthwhile pausing to reflect on the motivation behind a change of variables. This may help to guide the practitioner to choose the most suitable variables. Ultimately, we invoke a change of variables to simplify the evaluation of a multiple integral. One therefore chooses a transformation

$$(x(u,v), y(u,v)) : (x,y) \longmapsto (u,v)$$

to *either*

(a) transform region $D$ into a simpler region $E$ (in the Example 4.4 $E$ was a simple rectangle); *or*

(b) transform the integrand $f(x,y)$ into some simpler $F(u,v)$.

Thus, in the case of (a), one is guided by the shape of the region $D$: What mathematical expressions determine the boundary, and can these be used to define the new boundary?

In the case of (b), is the form of $f(x,y)$ suggestive of a suitable transformation? For example, $f(x,y) = g(x^2 + y^2)$ suggests the polar coordinate transformation $x = r\cos\theta$, $y = r\sin\theta$, so that $x^2 + y^2 = r^2$.

In both cases always look for symmetry features. This having been said, there will always be consequences.

(c) The region can be transformed into a more complicated one, even if the integrand becomes simpler;

(d) The integrand may become more complex, even if the transformed region is simpler (recall that a Jacobian for the transformation needs to be considered);

(e) Sometimes, however, we get lucky and *both* the function *and* the region become simpler.

## ✍ Mastery Check 4.9:

What is the image $R_2$ of the region $R_1$ bounded by the curves $x = y^2$, $y = \dfrac{1}{x}$, $y = \dfrac{2}{x}$, $x = \dfrac{y^2}{2}$, under the transformation $u = \dfrac{x}{y^2}$, $v = xy$?

Hint: Draw a picture for the region $R_1$ in the $xy$-plane and another for the image $R_2$ in the $uv$-plane. Then stop to think: Is this OK?

✍

### ✍  Mastery Check 4.10:

Calculate the volume of the solid defined by the intersection of the regions

$$x \geq 0, \quad z \geq 0, \quad z \leq x^2 + y^2, \quad 1 \leq x^2 + y^2 \leq 4.$$

Hint: Draw the graph of the solid (using MATLAB® is best). Sketch the region of integration in the $xy$-plane. Make an attempt at the integration in the $x, y$ variables. Then think about a suitable transformation.

✍

### ✍  Mastery Check 4.11:

This is an exercise with an ulterior motive — a bit like the last one, but more so!

Evaluate the double integral    $I = \displaystyle\iint_T e^{(y-x)/(y+x)} \, dA$

where $T$ is the triangle with corners at the points $(0,0)$, $(a,0)$, $(0,a)$.

Hint: An attempt to integrate in the $xy$-plane is likely to fail, but try anyway. Then make the simplest, most obvious, transformation: it works beautifully!

Finally, try the next most "obvious" transformation, polar coordinates. That works, too. You may need to recall a couple of trigonometric relations:

$$\tan(\theta - \pi/4) = \frac{\tan\theta - 1}{1 + \tan\theta}, \quad \cos(\theta - \pi/4) = \frac{\cos\theta + \sin\theta}{\sqrt{2}}.$$

✍

## 4.F  Triple integrals

To cement the ideas we've just introduced we now illustrate the case for functions of *three* variables integrated over regions of $\mathbb{R}^3$.

Suppose $f : \mathbb{R}^3 \longrightarrow \mathbb{R}$ is a continuous function, defined (at least) over a rectangular box $B$: $B = \{(x,y,z) : a_1 \leq x \leq b_1, \ a_2 \leq y \leq b_2, \ a_3 \leq z \leq b_3\}$.
          — we assume $B$ to be closed and bounded, $a_i \leq b_i < \infty$.

$B$ is shown in Figure 4.20.

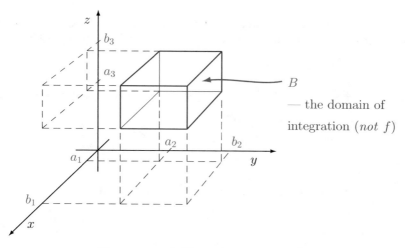

**Figure 4.20**  Box domain $B$ in $\mathbb{R}^3$.

The definition of a triple integral follows in analogy with those of double and single integrals. We just outline the relevant steps leading to the definition.

- Partition $B$ into small rectangular blocks $B_{ijk}$ of volumes
  $$\Delta V_{ijk} = \Delta x_i \Delta y_j \Delta z_k$$

- If $f \geq 0$ for all $\boldsymbol{x} \in B$, then we interpret the quantity $f$ as a *density* so that in choosing $(\xi_i, \eta_j, \zeta_k) \in B_{ijk}$, $f(\xi_i, \eta_j, \zeta_k)\Delta V_{ijk}$ will be an approximation to the mass of block $B_{ijk}$.

- The Riemann sum of all such masses in the partition,

$$\sigma_{nml} = \sum_{i=1}^{n}\sum_{j=1}^{m}\sum_{k=1}^{\ell} f(\xi_i, \eta_j, \zeta_k)\Delta V_{ijk},$$

  is an approximation to the total *mass* of the entire $B$.

- We then take the combined limits of the number of boxes to infinity with vanishing volumes. We therefore arrive at

**Definition 4.4**
*The **triple integral** of $f$ over $B$ is defined as the limit (if it exists)*

$$\lim_{\substack{n,m,\ell \to \infty \\ \max \sqrt{\Delta x_i^2 + \Delta y_j^2 + \Delta z_k^2} \to 0}} \sigma_{nm\ell} = \iiint_B f(x, y, z)\, \mathrm{d}V.$$

For regions of more general shape, starting with bounded regions, $S$, we can extend the definition of the integral of $f$ over $S \subset \mathbb{R}^3$ in analogy with the 2D version:

- we enclose $S$ in a closed and bounded box $B$: $S \subset B \subset \mathbb{R}^3$, and

- define $\hat{f}(x, y, z) = \begin{cases} f(x, y, z), & \boldsymbol{x} \in S \\ 0, & \boldsymbol{x} \notin S \end{cases}$,

we then have

$$\iiint_S f(x, y, z)\, \mathrm{d}V \equiv \iiint_B \hat{f}(x, y, z)\, \mathrm{d}V.$$

This now sets the stage for Section 4.G where the practical evaluation of triple integrals over general regions is discussed. In the meantime we have, as in the 1D and 2D cases, the following useful theorem and its corollaries.

**An important theorem and some corollaries — 3D version.**

**Theorem 4.1** *All continuous functions are integrable over compact subsets of their domains.*

**Corollary 4.1.1** *If $f \geq 0$, then*
$\iiint_B f\, \mathrm{d}V$ *is the "volume" of a 4-dimensional "solid" "under" $f$ "over" $B$.*

$\qquad\qquad\qquad\qquad\qquad\qquad$ — *not a very helpful interpretation.*

**Corollary 4.1.2**
*If $f \geq 0$ is a $\left.\begin{matrix} mass \\ charge \end{matrix}\right\}$ density, then $\iiint_B f\, \mathrm{d}V = \begin{cases} total\ mass \\ total\ charge \end{cases}$.*

$\qquad\qquad\qquad\qquad\qquad\qquad$ — *a more helpful interpretation.*

**Corollary 4.1.3** *If $f = 1$, then $\iiint_B 1\, \mathrm{d}V = volume\ of\ solid\ B$.*

$\qquad\qquad$ — *even more useful, especially for more general regions.*

**Corollary 4.1.4** *Average of $f(x, y, z)$ over $B = \dfrac{\iiint_B f\, \mathrm{d}V}{\iiint_B 1\, \mathrm{d}V} = \dfrac{\iiint_B f\, \mathrm{d}V}{vol. B}$.*

$\qquad\qquad\qquad\qquad$ — *this is true even for more complex regions.*

**Corollary 4.1.5** *Linearity: If $a, b \in \mathbb{R}$ then*

$$\iiint_B (af + bg)\, dV = a \iiint_B f\, dV + b \iiint_B g\, dV.$$

**Corollary 4.1.6** *Additivity: (very important)*
*If $B_1 \cap B_2 = \{\} \equiv \emptyset$, then*

$$\iiint_{B_1 \cup B_2} f\, dV = \iiint_{B_1} f\, dV + \iiint_{B_2} f\, dV.$$

As for the actual calculation of triple integrals the next section extends the ideas outlined in Section 4.C.

# 4.G   Iterated integration in $\mathbb{R}^3$

In evaluating a triple integral, we again make use of the idea of "slicing". The combination of a 3D integration domain and a function $f(x, y, z)$ results in a 4D graph, but we can visualize only the 3D domain. Keep in mind therefore that the first slice through the domain actually results in a *projection* of the graph of $f$ onto 3D.

In Figure 4.21 we first take horizontal slices through the 3D integration domain (left panel) which results in 2D regions (right panel). We then take constant $x$- or constant $y$-slices through the horizontal $z$-slice.

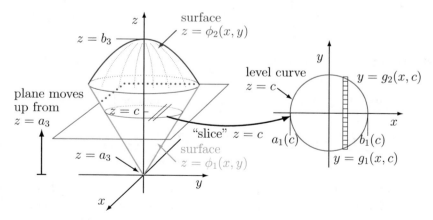

**Figure 4.21** The process of domain slicing: first in $z$, then in $x$, last in $y$.

In choosing the order in which we take slices we also commit ourselves to the order in which we perform the iterated integrals (in reverse). For instance, in the case of Figure 4.21 the iterated integral is the following:

$$\iiint_S f \, dV = \int_{a_3}^{b_3} dz \int_{a_1(z)}^{b_1(z)} dx \int_{g_1(x,z)}^{g_2(x,z)} f(x,y,z) \, dy.$$

That is, one integrates with respect to $y$ first, between limits that depend on $x$ and $z$, then one integrates with respect to $x$, between limits that depend on $z$. Finally, one integrates with respect to $z$ between two constants.

For the exact same problem we can consider vertical slices along the $x$-axis from $a_1$ to $b_1$. For each $x$ value we can take either $y$-slices or $z$-slices. Figure 4.22 shows the procedure corresponding to $x$-slices, then $y$-slices, and finally $z$-slices.

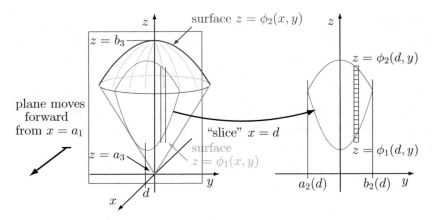

**Figure 4.22**  The process of domain slicing: first in $x$, then in $y$, last in $z$.

The interated integral for the particular case of Figure 4.22 is this

$$\iiint_S f \, dV = \int_{a_1}^{b_1} dx \int_{a_2(x)}^{b_2(x)} dy \int_{\phi_1(x,y)}^{\phi_2(x,y)} f(x,y,z) \, dz.$$

That is, one integrates with respect to $z$ first, between limits that depend on $x$ and $y$, then one integrates with respect to $y$, between limits that depend on $x$, and finally one integrates with respect to $x$ between two constants.

Generally, for any triple integral, the possible alternatives are these:

$$\iiint_S f\,\mathrm{d}V: \quad \text{$x$-slices} \quad \begin{cases} \displaystyle\int_{a_1}^{b_1} \mathrm{d}x \int_{a_2(x)}^{b_2(x)} \mathrm{d}y \int_{a_3(x,y)}^{b_3(x,y)} f\,\mathrm{d}z \\[2.5ex] \displaystyle\int_{a_1}^{b_1} \mathrm{d}x \int_{a_3(x)}^{b_3(x)} \mathrm{d}z \int_{a_2(x,z)}^{b_2(x,z)} f\,\mathrm{d}y \end{cases}$$

$$\text{$y$-slices} \quad \begin{cases} \displaystyle\int_{a_2}^{b_2} \mathrm{d}y \int_{a_1(y)}^{b_1(y)} \mathrm{d}x \int_{a_3(x,y)}^{b_3(x,y)} f\,\mathrm{d}z \\[2.5ex] \displaystyle\int_{a_2}^{b_2} \mathrm{d}y \int_{a_3(y)}^{b_3(y)} \mathrm{d}z \int_{a_1(y,z)}^{b_1(y,z)} f\,\mathrm{d}x \end{cases}$$

$$\text{$z$-slices} \quad \begin{cases} \displaystyle\int_{a_3}^{b_3} \mathrm{d}z \int_{a_1(z)}^{b_1(z)} \mathrm{d}x \int_{a_2(x,z)}^{b_2(x,z)} f\,\mathrm{d}y \\[2.5ex] \displaystyle\int_{a_3}^{b_3} \mathrm{d}z \int_{a_2(z)}^{b_2(z)} \mathrm{d}y \int_{a_1(y,z)}^{b_1(y,z)} f\,\mathrm{d}x \end{cases}$$

Thus, one triple integral gives rise to six possible iterated integrals. Note carefully the nature of the limits of each variable and how their dependencies vary from one integral to the next.

■ **Example 4.5:**

Evaluate the integral $\displaystyle\iiint_D xy\,\mathrm{d}V$, where $D$ is the interior of the sphere $x^2 + y^2 + z^2 = 1$ in the first octant, $0 \leq x, y, z \leq 1$.

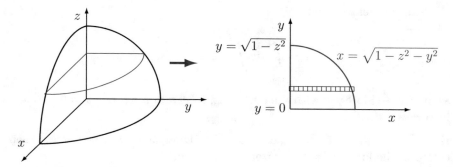

**Figure 4.23** The domain $D$, and the projection of a $z$-slice onto the $xy$-plane.

**Solution:** As sketched in Figure 4.23, we will take horizontal slices, $z = $ a constant, for $0 \le z \le 1$, and (for fun) integrate w.r.t. $x$ before we integrate w.r.t. $y$. The slices are $x^2 + y^2 \le 1 - z^2$, $x \ge 0$, $y \ge 0$, for a given $z$.

We need the bounds for the $x$-integration as functions of $y$.

These are $x = 0, \sqrt{1 - z^2 - y^2}$.

The bounds for the $y$-integration are $0, \sqrt{1 - z^2}$, and for the $z$-integration they are $0, 1$. So we get

$$
\iiint_D xy \, dV = \int_0^1 dz \int_0^{\sqrt{1-z^2}} dy \int_0^{\sqrt{1-z^2-y^2}} xy \, dx
$$

$$
= \int_0^1 dz \int_0^{\sqrt{1-z^2}} y \left[ \frac{1}{2} x^2 \right]_{x=0}^{\sqrt{1-z^2-y^2}} dy
$$

$$
= \int_0^1 dz \int_0^{\sqrt{1-z^2}} \frac{1}{2} y \left(1 - z^2 - y^2\right) dy
$$

$$
= -\int_0^1 \frac{1}{8} \left[ \left(1 - z^2 - y^2\right)^2 \right]_{y=0}^{\sqrt{1-z^2}} dz
$$

$$
= -\int_0^1 \frac{1}{8} \left(0 - \left(1 - z^2\right)^2\right) dz \; = \; \frac{1}{15}.
$$

■

✍  **Mastery Check 4.12:**

Rewrite the iterated integral

$$
I = \int_0^1 dz \int_z^1 dx \int_0^{x-z} f \, dy
$$

as an iterated integral with the outermost integration w.r.t. $x$ and innermost integration w.r.t. $z$.

Hint: Use the given limits to determine the 3D region of integration and then establish the limits of the new iterated integral.

✍

✍    **Mastery Check 4.13:**

Let $S$ be the volume of the body bounded by the planes $x = 0$, $y = 0$, and $z = 4$, and the surface $z = x^2 + y^2$. Calculate $I = \iiint_S x\,\mathrm{d}V$.    ✍

✍    **Mastery Check 4.14:**

Compute the volume of that part of the cylinder $x^2 - 2x + y^2 = 0$ cut off by the cylinder $z^2 - 2x = 0$. The region defined by this intersection is shown in Figure 4.24. (See also Figure 1.34 on Page 46.)

✍

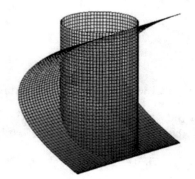

**Figure 4.24**  Two intersecting cylinders.

## 4.H  Change of variables in $\mathbb{R}^3$

Consider the triple integral of a continuous function over a closed and bounded domain,

$$I = \iiint_S f(x, y, z)\,\mathrm{d}V,$$

and the bijective $C^1$-transformation:

$$\tau : \mathbb{R}^3 \longrightarrow \mathbb{R}^3; \quad \boldsymbol{u} \mapsto \boldsymbol{x}(\boldsymbol{u}).$$

This mapping transforms an element of volume in the $xyz$-domain to a volume element in the $uvw$-domain, as suggested graphically in Figure 4.25.

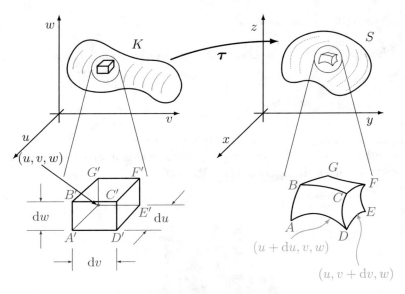

**Figure 4.25** A geometrical view of a change of variables in a 3D domain.

If $\mathrm{d}u, \mathrm{d}v, \mathrm{d}w \ll 1$ then (according to Section 1.A) the volume element in $S$ would be given by the absolute value of the scalar triple product

$$
\begin{aligned}
\mathrm{d}V &= \left| (\overrightarrow{BA} \times \overrightarrow{BC}) \cdot \overrightarrow{BG} \right| \\
&= \left| ((\mathrm{d}x_u \mathbf{e}_1 + \mathrm{d}y_u \mathbf{e}_2 + \mathrm{d}z_u \mathbf{e}_3) \times (\mathrm{d}x_v \mathbf{e}_1 + \mathrm{d}y_v \mathbf{e}_2 + \mathrm{d}z_v \mathbf{e}_3)) \right. \\
&\qquad\qquad \left. \cdot (\mathrm{d}x_w \mathbf{e}_1 + \mathrm{d}y_u \mathbf{e}_2 + \mathrm{d}z_w \mathbf{e}_3) \right|
\end{aligned}
$$

Invoking the chain rule, the scalar triple product can be written in determinant form (Section 1.A, Page 8).

$$
\mathrm{d}V = \left\| \begin{matrix}
\dfrac{\partial x}{\partial u}\mathrm{d}u & \dfrac{\partial y}{\partial u}\mathrm{d}u & \dfrac{\partial z}{\partial u}\mathrm{d}u \\[2mm]
\dfrac{\partial x}{\partial v}\mathrm{d}v & \dfrac{\partial y}{\partial v}\mathrm{d}v & \dfrac{\partial z}{\partial v}\mathrm{d}v \\[2mm]
\dfrac{\partial x}{\partial w}\mathrm{d}w & \dfrac{\partial y}{\partial w}\mathrm{d}w & \dfrac{\partial z}{\partial w}\mathrm{d}w
\end{matrix} \right\| = \left| \dfrac{\partial(x,y,z)}{\partial(u,v,w)} \right| \mathrm{d}u\,\mathrm{d}v\,\mathrm{d}w.
$$

As we found in the case of a change of variables in a double integral, a change of variables in a triple integral involves the Jacobian determinant for that transformation. The *absolute value* of the Jacobian is the scaling factor between the volume elements $\mathrm{d}V = \mathrm{d}x\,\mathrm{d}y\,\mathrm{d}z$ in $xyz$-space and $\mathrm{d}V' = \mathrm{d}u\,\mathrm{d}v\,\mathrm{d}w$ in $uvw$-space.

Consequently, we have the end result that

$$\iiint_S f(x,y,z)\,\mathrm{d}V = \iiint_K F(u,v,w)\left|\frac{\partial(x,y,z)}{\partial(u,v,w)}\right|\mathrm{d}V',$$

where $F(u,v,w) = f(x(u,v,w), y(u,v,w), z(u,v,w))$ and $S$ is the image of $K$ under $\tau$.

### ✍ Mastery Check 4.15:

Transform the iterated integral below into iterated integrals w.r.t. cylindrical coordinates, and w.r.t. spherical coordinates:

$$I = \int_0^1 \mathrm{d}x \int_0^{\sqrt{1-x^2}} \mathrm{d}y \int_0^{1+x+y} (x^2 - y^2)\,\mathrm{d}z.$$

(Do not proceed to evaluate the integrals. If you skip to the very last note in this chapter, you will see why!)

✍

## 4.I   $n$-tuple integrals

As we did in going from double integrals to triple integrals, all the preceding ideas, concepts and mathematical arguments can be generalized to $n$ dimensions.

#### $n$-tuple integrals

Suppose $S \subset \mathbb{R}^n$ is closed and bounded and we have $f : S \longrightarrow \mathbb{R}$ (the graph of $f \subset \mathbb{R}^{n+1}$).

Enclose $S$ in an $n$-dimensional box

$$[a_1, b_1] \times [a_2, b_2] \times \ldots \times [a_n, b_n].$$

Partition the box into $n$-dimensional boxes of size

$$\Delta x_1 \times \Delta x_2 \times \Delta x_3 \times \ldots \times \Delta x_n.$$

Choose                    $\xi_i \in [x_i, x_i + \Delta x_i] \quad i = 1, 2, \ldots, n.$

Form the sum

$$\sum \sum \cdots \sum f(\xi_1, \xi_2, \ldots, \xi_n)\Delta x_1 \Delta x_2 \ldots \Delta x_n.$$

If the limit of this sum as $n \to \infty$ and $|\Delta x| \to 0$ exists we call it the $n$-dimensional integral of $f$ over $S$

$$\iint \cdots \int_S f(x_1, x_2, \ldots, x_n) \, \mathrm{d}V_n = I,$$

where $\mathrm{d}V_n$ is an $n$-dimensional volume element.

**Iterated integrals**

If $S$ can be described as

$$S = \big\{ x = (x_1, x_2, \ldots, x_n) : (x_1, x_2, \ldots, x_{n-1}) \in W \subset \mathbb{R}^{n-1}$$
$$\text{and } \phi_1(x_1, x_2, \ldots, x_{n-1}) \le x_n \le \phi_2(x_1, x_2, \ldots, x_{n-1}) \big\}$$

then    $I = \displaystyle\iint \cdots \int_W \mathrm{d}x_1 \ldots \mathrm{d}x_{n-1} \int_{\phi_1}^{\phi_2} f(x) \, \mathrm{d}x_n = \cdots$

and, in fact,

$$I = \int_{\alpha_1}^{\alpha_2} \mathrm{d}x_1 \int_{\beta_1(x_1)}^{\beta_2(x_1)} \mathrm{d}x_2 \int_{\gamma_1(x_1,x_2)}^{\gamma_2(x_1,x_2)} \mathrm{d}x_3 \cdots \int_{\phi_1(x_1,\ldots,x_{n-1})}^{\phi_2(x_1,\ldots,x_{n-1})} f(x) \, \mathrm{d}x_n,$$

which is just one of $n!$ alternative iterated integrals.

**Change of variables**

Consider a bijective $C^1$ transformation: $\tau : u \longmapsto x(u)$, where the functions $x_i = x_i(u_1, u_2, \ldots, u_n)$, $i = 1, 2, \ldots, n$, are such that the Jacobian (Section 2.H)

$$J = \frac{\partial(x_1, \ldots, x_n)}{\partial(u_1, \ldots, u_n)} \ne 0.$$

The $n$-tuple integral $I$ of $f(x_1, \ldots, x_n)$ over $S$ is equal to the $n$-tuple integral of the product of $|J|$ and

$$F(u_1, \ldots, u_n) = f\big(x_1(u_1, u_2, \ldots, u_n), \ldots, x_n(u_1, u_2, \ldots, u_n)\big)$$

over the pre-image $E$ of $S$ under $\tau$:

$$I = \iint \cdots \int_S f(x) \mathrm{d}V_n = \iint \cdots \int_E F(u)|J| \mathrm{d}V_n'.$$

See Section 5.A for an expansion of $J$.

# 4.J   Epilogue: Some practical tips for evaluating integrals

Recall that ...

> * If $f(-x) = -f(x)$, then $f$ is *odd*, and $\displaystyle\int_{-a}^{a} f(x)\,dx = 0$.
>
> $\qquad\qquad\qquad$ — for example: $\sin(x)$, $x^3$, $x^7$, $\arctan(x)$
>
> * If $f(-x) = f(x)$, then $f$ is *even*, and $\displaystyle\int_{-a}^{a} f(x)\,dx = 2\int_{0}^{a} f(x)\,dx$.
>
> $\qquad\qquad\qquad$ — for example: $\cos(x)$, $x^2$, $\sin^2(x)$

Did you know that for functions of $n \geq 2$ variables there are other possible symmetry features?

> * If $f(-x, y) = -f(x, y)$, then $f$ is *odd with respect to* $x$, which means that
>
> $$\int_{c}^{d} dy \int_{-a}^{a} f(x, y)\,dx = 0. \qquad\text{— for example: } xy^2,\ \sin(x).y,\ \arctan(xy)$$
>
> * If $f(-x, y) = f(x, y)$, then $f$ is *even with respect to* $x$, which means that
>
> $$\int_{c}^{d} dy \int_{-a}^{a} f(x, y)\,dx = 2\int_{c}^{d} dy \int_{0}^{a} f(x, y)\,dx.$$
>
> $\qquad\qquad\qquad$ — for example: $x^2 y$, $\cos(x).y^3$, $\arctan(x^2 y)$
>
> * Similarly, we may have functions $f(x, y)$ which are *odd or even with respect to* $y$.
>
> * Now, for $f : \mathbb{R}^2 \longrightarrow \mathbb{R}$ we have a *new* possibility, indicated in Figure 4.26:
>
> If $f(x, y) = f(y, x)$, then $f$ is *symmetric* across $y = x$.
> If $f(x, y) = -f(y, x)$, then $f$ is *antisymmetric* across $y = x$.

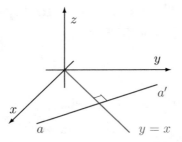

**Figure 4.26** The oblique symmetry line $y = x$.

Some examples of functions *symmetric* from point $a$ to point $a'$ are
$$f(x, y) = x + y,$$
$$f(x, y) = x^2 + y^2,$$
$$f(x, y) = 1/\sqrt{x^2 + y^2},$$
$$f(x, y) = b + (y - x)^2 \text{ (Figure 4.27(a)).}$$

Some examples of functions *antisymmetric* from point $a$ to point $a'$ are
$$f(x, y) = x - y,$$
$$f(x, y) = x^2 - y^2,$$
$$f(x, y) = \sin(y - x) \text{ (Figure 4.27(b)).}$$

(a)                                    (b)

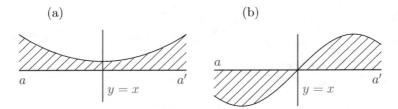

**Figure 4.27** (a) Cross-section through $f(x, y) = b + (y - x)^2$; (b) Cross-section through $f(x, y) = \sin(y - x)$.

# 4.K   Supplementary problems

### Section 4.B

1. Determine $\iint_D x \cos xy \, dx \, dy$,

   where $D = \{(x, y) \in \mathbb{R}^2 : 0 \leq x \leq \pi/2, \ 0 \leq y \leq 1\}$.

   (Hint: Treat this as an interated integral and integrate with respect to $y$ first.)

2. Determine $\iint_D 2xy \sec^2(x^2 y) \, dx \, dy$,

   where $D = \{(x, y) \in \mathbb{R}^2 : 0 \leq x \leq 1, \ 0 \leq y \leq \pi/4\}$.

### Section 4.C

3. Compute the integral $\iint_D \sqrt{xy} \, dx \, dy$,

   where $D = \{(x, y) : 0 \leq x \leq 1, \ x^3 \leq y \leq x^2\}$.

4. Compute the integral $\iint_D xy \, dx \, dy$,

   where $D = \{(x, y) : \cos y \leq x \leq \sin y, \ \dfrac{\pi}{4} \leq y \leq \dfrac{5\pi}{4}\}$.

### Section 4.D

5. Check whether the following integrals converge. If they do, compute them.

   (a) $\iint_D \dfrac{dx \, dy}{1 + (x + y)^2}$, $D = \{(x, y) : x > 0, y > 0\}$.

   (b) $\iint_D \dfrac{dx \, dy}{\sqrt{xy}}$, $D = \{(x, y) : 0 < x < 1, 0 < y < 1\}$.

   (c) $\iint_D xe^{-(y + x^2/y)} \, dx \, dy$, where $D$ is the first quadrant.

   (Hint: Consider the rectangle $\{(x, y) : 0 < x < A, 0 < y < B\}$, and let $A \to \infty$, $B \to \infty$.)

**Section** 4.E

6. Determine $\iint_D y e^{-(x^2+y^2)} \, dx \, dy$, where $D = \{(x,y) \in \mathbb{R}^2 : x^2 + y^2 \leq 1, y > 0\}$.

7. A device commonly used to determine $I = \int_{-\infty}^{\infty} e^{-x^2/2} \, dx$ is as follows:

   From symmetry (see Section 4.J) we may write
   $$I^2 = 4 \left( \int_0^{\infty} e^{-x^2/2} \, dx \right) \left( \int_0^{\infty} e^{-y^2/2} \, dy \right) = \iint_R e^{-(x^2+y^2)/2} \, dx \, dy,$$
   where $R$ is the first quadrant on the Cartesian plane.

   Use a polar coordinate transformation to evaluate this integral.

8. Let $D = \{(x,y) : x \geq 0, y \geq 0, x^3 + y^3 \leq 0\}$.

   By introducing a change of variables $(x,y) \to (u = x^3, v = y^3)$ evaluate the double integral
   $$\iint_D x^2 y^3 \sqrt{1 - x^3 - y^3} \, dA.$$
   Express your answer in terms of $\Gamma(1/3)$ and $\Gamma(1/6)$.

**Section** 4.F

9. Recall from elementary physics that

   a) If a force $F = mg$ (where $g$ is the acceleration due to gravity) acts on a mass $m$ at a horizontal distance $x$ from a (pivot) point $O$, then its *moment* about $O$ is $F.x$;
   b) If a mass $m$ moves so that its distance from a point $O$ is constrained to be $x$, then its *moment of inertia* about $O$ is $m.x^2$.

   Now consider the integral $\iiint_V f(x,y,z) \, dx \, dy \, dz$.
   Interpret this integral if

   (i) $f(x,y,z) = 1$.

   (ii) $f(x,y,z)$ is the material density at point $(x,y,z)$.

   (iii) $f(x,y,z) = \sqrt{x^2 + y^2} \times$ the material density at point $(x,y,z)$.

   (iv) $f(x,y,z) = (x^2 + y^2) \times$ the material density at point $(x,y,z)$.

**Section** 4.G

10. For each of the following, sketch the region and evaluate the integral

(i)   $\displaystyle\int_0^1 dz \int_0^z dy \int_0^y dx$

(ii)  $\displaystyle\int_0^2 dx \int_1^x dy \int_2^{x+y-1} y dz$

(iii) $\displaystyle\int_0^1 dx \int_x^{\sqrt{x}} dy \int_{1-x-y}^{1+x+y} xy dz$

11. Let $S$ be the solid in the first octant containing the origin and bounded by sections of the plane $y = 1 + 2x$ and the sphere $x^2 + y^2 + z^2 = 4$. For any integrable function $f$ defined on $S$ write the triple integral $\iiint_S f dV$ as an iterated integral with respect to $x, y, z$ in six different ways.

That is, you are to determine the respective limits depicted in the iterated integral formulae on Page 209.

12. Determine the volume shown in Figure 4.28 enclosed by the two surfaces

$$z = 8 - x^2 - y^2 \qquad \text{and} \qquad z = x^2 + 3y^2.$$

Hint: Set up the integral as $V = \displaystyle\int_a^b dx \int_{a_1(x)}^{a_2(x)} dy \int_{a_3(x,y)}^{a_4(x,y)} 1. dz$, that is, take horizontal $z$-slices.

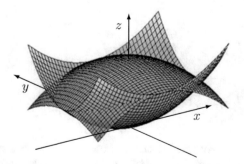

**Figure 4.28** Two intersecting paraboloids.

**Section** 4.H

13. Evaluate the triple integral of the function

$$f(x, y, z) = x^2 y^2 z$$

over the region $R$ bounded by the cone $x^2 + y^2 = xz$ and the planes $z = 0$ and $z = c$. (See Figure 4.29.)

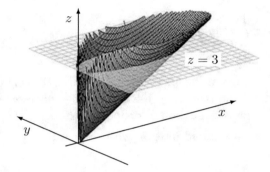

**Figure 4.29** The graph of $x^2 + y^2 = xz$ intersected by $z = 3$.

14. Show that

$$\int_0^\infty dx \int_0^\infty xye^{-(x^2+y^2+2xy\cos\alpha)}dy = \frac{\sin\alpha - \alpha\cos\alpha}{4\sin^3\alpha}$$

where $0 < \alpha < \pi$.

15. Transform each of the following two iterated integrals into iterated integrals with respect to both cylindrical and spherical coordinates.

(i) $\int_0^1 dx \int_0^{\sqrt{1-x^2}} dy \int_0^{1+x+y} dz$

(ii) $\int_{-1}^1 dx \int_{-\sqrt{1-x^2}}^{\sqrt{1-x^2}} dy \int_{\sqrt{x^2+y^2}}^1 dz$

16. Determine the volume of the region in the first octant bounded by the surface $x^4 + y^4 + z^4 = c^4$ and the coordinate planes. Express your answer in terms of a Gamma function.

17. Let $T$ be the tetrahedron with vertices $(a, 0, 0)$, $(0, a, 0)$, $(0, 0, a)$, and $(0, 0, 0)$. Show that

$$\iiint_T \frac{x^{1/2}y^{3/2}z^{5/2}}{(x + y + z)^{13/2}} \, dV = \frac{2\pi a}{3003}.$$

Hint: Find a transformation that maps $T$ to the unit cube. You may need to utilize the integral and numerical properties of Gamma functions.

## Section 4.I

18. Suppose $f(x, y, z, t)$ describes the rate of change of electric charge density at point $(x, y, z)$ at time $t$ throughout a volume $V$.

Give a meaning to the integral

$$I = \iiiint_D f(x, y, z, t) \, dx \, dy \, dz \, dt$$

where $D = \{(x, y, z, t) : x^2 + y^2 + z^2 \le a^2, \; 0 \le t \le T\}$ and indicate how the integration might be carried out.

Write down the result of the integration in the case that $f(x, y, z, t) = c$, a constant.

19. Show that

$$\int_0^x dx_1 \int_0^{x_1} dx_2 \int_0^{x_2} dx_3 \cdots \int_0^{x_{n-1}} f(x_n) dx_n$$
$$= \frac{1}{(n-1)!} \int_0^x (x - t)^{n-1} f(t) dt.$$

20. Devise suitable $n$-dimensional spherical polar coordinates to satisfy

$$x_1^2 + x_2^2 + \cdots + x_n^2 = a^2$$

and, using integral properties of Gamma functions, derive the volume of the $n$-ball.

21. Using the results of the foregoing problem, determine the volume of the $n$-dimensional ellipsoid:

$$\frac{(x_1 - b_1)^2}{a_1^2} + \frac{(x_2 - b_2)^2}{a_2^2} + \cdots + \frac{(x_n - b_n)^2}{a_n^2} \le 1.$$

**Section** 4.J

22. Give reasons why you may decide whether the following integrals are zero or not by inspection only, that is, without any computation.

(a) $\displaystyle\iint_D xe^{-x^2-y^2}\,dx\,dy$, where $D = \{(x,y) : |x| \le 1, |y| \le 1\}$.

(b) $\displaystyle\iint_D xe^{-x^2-y^2}\,dx\,dy$, where $D = \{(x,y) : |x| \le 1, 0 \le y \le 1\}$.

(c) $\displaystyle\iint_D ye^{-x^2-y^2}\,dx\,dy$, where $D = \{(x,y) : |x| \le 1, 0 \le y \le 1\}$.

(d) $\displaystyle\iint_D (x-y)e^{-x^2-y^2}\,dx\,dy$, where $D = \{(x,y) : |x| \le 1, |y| \le 1\}$.

(e) $\displaystyle\iint_D (x-y)e^{-x^2-y^2}\,dx\,dy$, where $D = \{(x,y) : |x| \le 1, 0 \le y \le 1\}$.

(f) $\displaystyle\iint_D (x-y)^2 e^{-x^2-y^2}\,dx\,dy$, where $D = \{(x,y) : |x| \le 1, |y| \le 1\}$.

(g) $\displaystyle\iint_D (x-y)^2 \sin(x-y)\,dx\,dy$, where $D = \{(x,y) : |x| \le 1, |y| \le 1\}$.

(h) $\displaystyle\iint_D (x-y)\sin(x-y)\,dx\,dy$, where $D = \{(x,y) : |x| \le 1, |y| \le 1\}$.

(i) $\displaystyle\iint_D (x-y)\sin(x+y)\,dx\,dy$, where $D = \{(x,y) : |x| \le 1, |y| \le 1\}$.

23. Let $S$ be the unit ball in $\mathbb{R}^3$. Show that $\displaystyle\iiint_S f(x,y,z)dV = -4\pi$, where
$$f(x,y,z) = -3 + 2y + (x^4 + y^6 + z^8)\sin x^3.$$

# Chapter 5

# Vector calculus

The majority of systems that arise in engineering and in the physical sciences fall into one of three camps: kinematic, dynamic, and static systems. Certainly in the first two, but even in the third camp, a system is only partially described by magnitudes of quantities. Systems in motion but also systems in a state of balance or equilibrium can only be completely characterized when directional dependencies are considered. The complete characterization of a system is therefore achieved by quantities that describe both direction and magnitude. These are vector-valued functions which vary with respect to specified independent variables. A force acting on an object is an example of a vector-valued function, as is the object's response in terms of its velocity of motion. Another example, this time of a distributed character, is the flow field of a fluid continuum.

This chapter brings together the concepts that were introduced in Chapters 2 and 4 for scalar-valued functions and extends their applications to functions that are themselves vectors. However, the result is not always a simple generalization from one to many dependent variables. We will discover a range of new results, new concepts, and new features, which hold specifically for vector-valued functions.

## 5.A  Vector-valued functions

We have already named some examples of vector-valued functions, force and velocity, but it helps to have a more general definition that is not tied to a specific application.

© Springer Nature Switzerland AG 2020
S. J. Miklavcic, *An Illustrative Guide to Multivariable and Vector Calculus*,
https://doi.org/10.1007/978-3-030-33459-8_5

**Definition 5.1**

*A vector,* $\boldsymbol{f}$*, is called a vector-valued (m-dimensional) function of the vector variable* $\boldsymbol{x}$ *if each of its m components,* $f_i$*, are real-valued functions of* $\boldsymbol{x}$*:*

$$f_i = f_i(\boldsymbol{x}); \quad f_i : \mathbb{R}^n \longrightarrow \mathbb{R}, \ i = 1, 2, \ldots, m$$

*and*

$$\boldsymbol{f}(\boldsymbol{x}) = \big(f_1(\boldsymbol{x}), f_2(\boldsymbol{x}), \ldots, f_m(\boldsymbol{x})\big).$$

We say that $\quad \boldsymbol{f} : \mathbb{R}^n \longrightarrow \mathbb{R}^m \quad$ defines a transformation $\quad \boldsymbol{x} \longmapsto \boldsymbol{y} = \boldsymbol{f}(\boldsymbol{x})$:

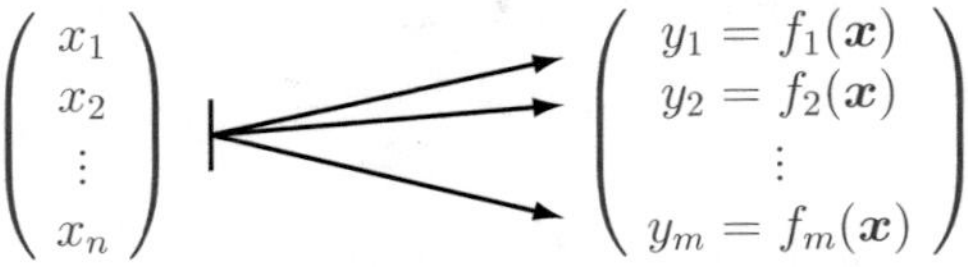

$$\begin{pmatrix} x_1 \\ x_2 \\ \vdots \\ x_n \end{pmatrix} \longmapsto \begin{pmatrix} y_1 = f_1(\boldsymbol{x}) \\ y_2 = f_2(\boldsymbol{x}) \\ \vdots \\ y_m = f_m(\boldsymbol{x}) \end{pmatrix}$$

**Figure 5.1** The vector mapping $\boldsymbol{x} \longmapsto \boldsymbol{y} = \boldsymbol{f}(\boldsymbol{x})$.

The $f_i$, $i = 1, 2, \ldots, m$, in Figure 5.1 are the components of $\boldsymbol{f}$ in the corresponding orthogonal directions $\mathbf{e}_i$, $i = 1, 2, \ldots, m$ in $\mathbb{R}^m$ (that is, $\mathbf{i}, \mathbf{j}, \mathbf{k}$ in $\mathbb{R}^3$).

From this most general form we now consider three important specific classes of vector-valued functions.

**I. Curves** $\boldsymbol{f} : \mathbb{R} \to \mathbb{R}^m$

The dependence here is on a single variable. For applications in physics where we use $(x, y, z)$ to denote position in $\mathbb{R}^3$, we denote the independent variable by $t$. The vector function in general defines a transformation $t \longmapsto \boldsymbol{f}(t)$ to a point in $\mathbb{R}^m$. As the independent variable varies over an interval domain $I$ in $\mathbb{R}$, the point traces out a *curve* in $\mathbb{R}^m$, as illustrated in Figure 5.2 for the case $m = 2$.

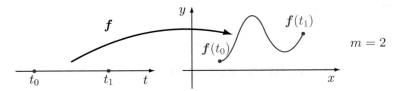

**Figure 5.2** A curve in 2-space.

Physically, this mapping describes, for example, the *path* or *trajectory* of a particle in motion. It is the foundation stone of the field of kinematics.

**Figure 5.3**  The trajectory in $\mathbb{R}^3$.

■  **Example 5.1:**

A ball rolling down a funnel at uniform vertical speed, Figure 5.3. The position $r$ of the ball is time-dependent ($t$-dependent).

$$x = \left(1 - \frac{t}{4\pi}\right)^2 \cos t$$

$$y = \left(1 - \frac{t}{4\pi}\right)^2 \sin t$$

$$z = 1 - \frac{t}{4\pi}$$

$$t = [0, 4\pi]$$

■

From the specific perspective of particle trajectories it is usual to replace $f$ with $r$ to reflect the application of vector functions to position in space. We are thus motivated to write

$$r(t) = \big(x_1(t), x_2(t), \ldots, x_m(t)\big)$$

to represent a curve in $m$-space. In $\mathbb{R}^3$ we use the notation

$$r(t) = \big(x(t), y(t), z(t)\big).$$

The physical world places some restrictions on the types of vector functions that can be used to describe curves in space. The primary restrictions relate to continuity and differentiability.

**Theorem 5.1**

*Suppose $I$ is an open connected interval in $\mathbb{R}$. The vector $\boldsymbol{r}(t)$ is a continuous vector function of $t \in I$, with continuous first derivatives (that is, is a $C^1$ function) if all of its component functions are $C^1$ functions of $t \in I$.*

As with all derivatives, the derivative vector is defined in terms of a converged limiting process, much the same as described in Chapter 1, but now applied to each component of $\boldsymbol{r}(t)$, and the results are recombined.

The preceding definition means that

$$\lim_{\Delta t \to 0} \frac{\boldsymbol{r}(t + \Delta t) - \boldsymbol{r}(t)}{\Delta t}$$

$$= \left( \lim_{\Delta t \to 0} \frac{x_1(t + \Delta t) - x_1(t)}{\Delta t}, \ldots, \lim_{\Delta t \to 0} \frac{x_m(t + \Delta t) - x_m(t)}{\Delta t} \right)$$

$$= \left( \frac{\mathrm{d}x_1}{\mathrm{d}t}(t), \ldots, \frac{\mathrm{d}x_m}{\mathrm{d}t}(t) \right) \quad \text{— if all } x_i \text{ are } C^1$$

$$\equiv \underbrace{\frac{\mathrm{d}\boldsymbol{r}}{\mathrm{d}t}(t)}_{\text{definition}} = \underbrace{\boldsymbol{v}(t)}_{\text{connection to physics — the velocity of particle motion}} = \boldsymbol{r}'(t).$$

Some finite steps in this limit process are shown in Figure 5.4.

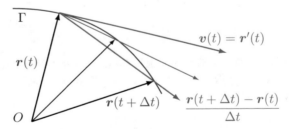

**Figure 5.4** The limit process for the curve $\Gamma$ in $\mathbb{R}^n$.

**Rules for differentiating curve vectors**

With the definition of the derivative of a vector-valued function of one variable being a generalization of the derivative of a scalar function of one variable, it is natural to expect that the rules for differentiating scalar functions also generalize. These can be proved by breaking the vector functions down into their components and applying concepts from single-variable calculus to each component, the results are combined thereafter.

For differentiable vector and scalar functions

$$\boldsymbol{u}, \boldsymbol{v} : \mathbb{R} \longrightarrow \mathbb{R}^n, \quad \phi : \mathbb{R} \longrightarrow \mathbb{R}, \quad f : \mathbb{R}^n \longrightarrow \mathbb{R},$$

it can be shown that

(a) $\dfrac{\mathrm{d}}{\mathrm{d}t}\Big(\boldsymbol{u}(t) + \boldsymbol{v}(t)\Big) \;=\; \boldsymbol{u}'(t) + \boldsymbol{v}'(t)$

(b) $\dfrac{\mathrm{d}}{\mathrm{d}t}\Big(\phi(t)\boldsymbol{u}(t)\Big) \;=\; \dfrac{\mathrm{d}}{\mathrm{d}t}\Big(\phi(t)u_1(t) + \cdots + \phi(t)u_m(t)\Big)$
$$\qquad\qquad\qquad\;=\; \phi'(t)\boldsymbol{u}(t) + \phi(t)\boldsymbol{u}'(t)$$

(c) $\dfrac{\mathrm{d}}{\mathrm{d}t}\Big(\boldsymbol{u}(t) \cdot \boldsymbol{v}(t)\Big) \;=\; \boldsymbol{u}'(t) \cdot \boldsymbol{v}(t) + \boldsymbol{u}(t) \cdot \boldsymbol{v}'(t)$ — the scalar product rule

(d) $\dfrac{\mathrm{d}}{\mathrm{d}t}\Big(f\big(\boldsymbol{u}(t)\big)\Big) \;=\; \boldsymbol{\nabla}f\big(\boldsymbol{u}(t)\big) \cdot \boldsymbol{u}'(t)$ — the chain rule

(e) $\dfrac{\mathrm{d}}{\mathrm{d}t}\Big(\boldsymbol{u}\big(\phi(t)\big)\Big) \;=\; \boldsymbol{u}'\big(\phi(t)\big)\phi'(t)$ — another chain rule application

and, for $n = 3$, we also have

(f) $\dfrac{\mathrm{d}}{\mathrm{d}t}\Big(\boldsymbol{u}(t) \times \boldsymbol{v}(t)\Big) \;=\; \boldsymbol{u}'(t) \times \boldsymbol{v}(t) + \boldsymbol{u}(t) \times \boldsymbol{v}'(t)$
— the vector product rule — the order of the vectors is important

✍  **Mastery Check 5.1:**
Prove the derivative laws (c)–(f).

✍

**Remark**

* Within the context of particle mechanics one can make a number of other associations, this time for the derivatives of $\boldsymbol{r}(t)$:

| | |
|---|---|
| $\boldsymbol{r}(t)$ | — particle position at time $t$ |
| $\boldsymbol{r}'(t) = \boldsymbol{v}(t)$ | — particle velocity at time $t$ |
| $\boldsymbol{r}''(t) = \boldsymbol{a}(t)$ | — particle acceleration at $t$ |
| $|\boldsymbol{v}(t)| = v(t) = \sqrt{v_1^2(t) + \cdots + v_m^2(t)}$ | — particle speed at time $t$ |

## Elementary differential and integral geometry of curves

Describing trajectories, which are fundamental to particle mechanics, is one application of a vector-valued function of one variable. In that case, the physical interest is divided between determining where the particle is at any point in time and computing the particle's position history, *i.e.* the entire path the particle took leading up to that point. Another area of interest focuses on the local properties of the path itself, as a turning and twisting curve in space. While this perspective can potentially increase our understanding of particle dynamics, its real value lies in its application to differential geometry, continuum mechanics of materials (fluid and solid) and general relativity. The fundamental property on which this perspective is based is that of the *tangent vector*.

In order for the tangent vector to be well defined at any point along the curve, the curve must be such that the limit process defining the derivative can be executed (continuity) and then for that limit to exist (differentiability). Beyond these conditions, which are applied in single-variable calculus, we have the further condition that the derivative of the vector is nonzero. This leads to the concept of curve *smoothness*.

---

**Definition 5.2**
*Suppose $I$ is an open connected interval in $\mathbb{R}$. A curve $\Gamma$, described by a vector function $\boldsymbol{r}(t)$ for $t \in I$, is called* **smooth** *if $\boldsymbol{r}(t)$ is a $C^1$-function $\left( \dfrac{\mathrm{d}\boldsymbol{r}}{\mathrm{d}t} \ exists \right)$ and $\dfrac{\mathrm{d}\boldsymbol{r}}{\mathrm{d}t}$* **never** *vanishes for any $t \in I$.* $\left[ \dfrac{\mathrm{d}\boldsymbol{r}}{\mathrm{d}t} = \boldsymbol{v} \neq 0. \right]$

---

## Remarks

* The condition $\boldsymbol{r}'(t) \neq 0$ means that not all components of the vector $\boldsymbol{r}'(t)$ can vanish simultaneously. It may still arise that one or more components vanish, but not all $n$ at the same $t$ value.

* In many texts, points $\boldsymbol{r}(t)$ for $t \in I$ where $\boldsymbol{r}'(t)$ exists and is nonzero are also called *regular points*. A curve possessing only regular points is therefore also called *regular*.

> **Definition 5.3**
> *Suppose $I$ is an open connected interval in $\mathbb{R}$. For a smooth curve $\Gamma$ defined by $\boldsymbol{r}(t)$ for $t \in I$, the vector function $\boldsymbol{v}(t) = \boldsymbol{r}'(t)$ is a **tangent vector** to $\Gamma$ at the point $\boldsymbol{r}(t)$.*

The tangent vector is the end result of the limit process illustrated schematically in Figure 5.4. While it already provides new information about the curve (indicating the direction the curve continues with increasing $t$) much more can be derived from it.

> **Definition 5.4**
> *For a smooth curve $\Gamma$ described by a vector function $\boldsymbol{r}(t)$ defined on an open connected interval $I$, let $\boldsymbol{T}(t) = \dfrac{\boldsymbol{r}'(t)}{|\boldsymbol{r}'(t)|}$ be the unit vector in the direction of the tangent vector $\boldsymbol{r}'(t)$.*
>
> *Then at those points of $\Gamma$ for which $\boldsymbol{T}'(t) \neq 0$ we define the **unit principal normal vector** to $\Gamma$ as $\boldsymbol{N}(t) = \dfrac{\boldsymbol{T}'(t)}{|\boldsymbol{T}'(t)|}$, and we define the **binormal vector** to $\Gamma$ as $\boldsymbol{B}(t) = \boldsymbol{T}(t) \times \boldsymbol{N}(t)$.*

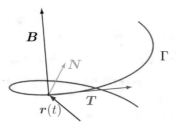

**Figure 5.5** Local orthogonal vectors $\boldsymbol{T}$, $\boldsymbol{N}$, and $\boldsymbol{B}$.

### ✍ Mastery Check 5.2:

Show that for any smooth curve in 3D, $\boldsymbol{T}(t)$ and $\boldsymbol{T}'(t)$ are orthogonal.

Hint: Differentiate $\boldsymbol{T} \cdot \boldsymbol{T}$ with respect to $t$ (see Rule (c) on Page 227).

What conclusions can you then draw about $\boldsymbol{N}(t)$ and $\boldsymbol{N}'(t)$, and $\boldsymbol{B}(t)$ and $\boldsymbol{B}'(t)$?

✍

**Remarks**

* For each $t$, the orthogonal vectors $\boldsymbol{T}(t)$ and $\boldsymbol{N}(t)$ define the *osculating plane* whose orientation (in 3D) changes with $t$. The binormal vector $\boldsymbol{B}(t)$ is orthogonal to the osculating plane.

* The vectors $\boldsymbol{T}$, $\boldsymbol{N}$, and $\boldsymbol{B}$, shown in Figure 5.5, define a local orthonormal set of vectors, much the same as the vectors $\mathbf{i}, \mathbf{j}, \mathbf{k}$. That is, they define a right-handed coordinate system. However, in contrast to the Cartesian unit vector set, the coordinate system moves and changes direction as $t$ increases. The system is appropriately called a *moving trihedral*. The vectors $\{\boldsymbol{T}, \boldsymbol{N}, \boldsymbol{B}\}$ establish what is called the *Frenet* framework.

* As a consequence of the foregoing remark, any 3D vector $\boldsymbol{u}(\boldsymbol{r}(t))$ relevant to the curve can be expressed in terms of the moving trihedral system,

$$\boldsymbol{u}(\boldsymbol{r}(t)) = \alpha(t)\boldsymbol{T}(t) + \beta(t)\boldsymbol{N}(t) + \gamma(t)\boldsymbol{B}(t).$$

In particular, this holds for $\boldsymbol{T}'(t)$, $\boldsymbol{N}'(t)$ and $\boldsymbol{B}'(t)$ as revealed by Definition 5.5 and Mastery Check 5.3.

---

**Definition 5.5**

*Let $\Gamma$ be a smooth curve described by $\boldsymbol{r}(t)$ defined on an open connected interval $I$. At any point $\boldsymbol{r}(t)$ on $\Gamma$ the non-negative function* $\kappa(t) = \dfrac{|\boldsymbol{T}'(t)|}{|\boldsymbol{r}'(t)|}$ *is called the* **curvature** *of $\Gamma$ at that point, and the real-valued function $\tau(t)$, defined such that $\boldsymbol{B}'(t) = -\tau(t)\,|\boldsymbol{r}'(t)|\,\boldsymbol{N}(t)$, is called the* **torsion** *of $\Gamma$ at that point.*

---

The function $\tau(t)$ gives a measure of the tendency of the curve to twist out of the osculating plane. It can be positive, zero, or negative. On the other hand, the curvature $\kappa$, which measures the extent to which the curve is bent at a point, is always non-negative ($\kappa(t) \geq 0$).

✍ **Mastery Check 5.3:**

Prove that $\boldsymbol{N}'(t) = -\kappa(t)\,|\boldsymbol{r}'(t)|\,\boldsymbol{T}(t) + \tau(t)\,|\boldsymbol{r}'(t)|\,\boldsymbol{B}(t)$.

Hint: Differentiate $\boldsymbol{N}(t) = \boldsymbol{B}(t) \times \boldsymbol{T}(t)$ with respect to $t$. See Rule (f) on Page 227.

✍

## ■   Example 5.2:

A straight line, $r(t) = r_0 + tu$, has a constant tangent vector, $T(t) = u/|u|$, and therefore has zero curvature, *i.e.* $\kappa(t) = 0$.

A circle in 2D, $r(\theta) = r_0 + a(\cos\theta, \sin\theta)$, or 3D,

$$r(\theta) = r_0 + a(\sin\phi\cos\theta, \sin\phi\sin\theta, \cos\phi) \quad \text{with} \quad \phi = f(\theta),$$

has constant curvature, $\kappa(\theta) = 1/a$, the reciprocal of the circle radius. The torsion also reduces to $\tau(\theta) = 0$ since the circle remains in the same plane. (For example, set $b = 0$ in Mastery Check 5.4. See also Supplementary problem 4.)                                                                          ■

## ✍   Mastery Check 5.4:

Consider the helical curve in 3D described by the vector function

$$r(t) = (a\cos t, a\sin t, bt),\ 0 \le t \le 2\pi.$$

Compute $r'(t)$, $r''(t)$, $T(t)$, $N(t)$, and $B(t)$.

Verify that $\kappa^2 = \dfrac{a^2}{(a^2+b^2)^2}$, and that $\tau(t) = \dfrac{b}{a^2+b^2}$ for all $t$.

✍

From the differential properties of curves we move now to integral properties starting with curve length. Not surprisingly, the length of a curve is also the total path length travelled by a particle along that curve. For a formal justification see Section 5.C.

---

**Definition 5.6**

*The **arc length** of a smooth curve $\Gamma$ described by $r(t)$ defined on an open connected interval $I$ measured from $t = t_0 \in I$ to an arbitrary $t \in I$ is*

$$s = \int_{t_0}^{t} v(\tau)\mathrm{d}\tau. \tag{5.1}$$

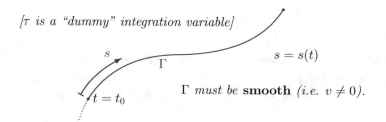

*[$\tau$ is a "dummy" integration variable]*

$s = s(t)$

$\Gamma$ *must be **smooth** (i.e. $v \ne 0$).*

**Figure 5.6** Arc length parameter, $s$.

---

By differentiating both sides of Equation (5.1) with respect to $t$, using Leibniz's rule (Page 99), we verify the fundamental theorem of calculus

$$\mathrm{d}s/\mathrm{d}t = |\boldsymbol{r}'(t)| = v(t) > 0.$$

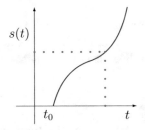

$s$ — a one-to-one and onto function of $t$, which means that $t$ is a function of $s$, that is, $t = t(s)$

**Figure 5.7**  Arc length is a one-to-one function of $t$.

Given the one-to-one relationship between $t$ and $s$ (Figures 5.6 and 5.7), we can parameterize $\Gamma$ with respect to $s$ instead of $t$. We then find that

$$\frac{\mathrm{d}\boldsymbol{r}}{\mathrm{d}s} = \frac{\mathrm{d}\boldsymbol{r}}{\mathrm{d}t}\frac{\mathrm{d}t}{\mathrm{d}s} = \frac{\boldsymbol{r}'(t)}{|\boldsymbol{r}'(t)|}$$

— *the unit tangent vector* to $\Gamma$ (normalized velocity)

which implies that

$$\left|\frac{\mathrm{d}\boldsymbol{r}}{\mathrm{d}s}\right| = 1.$$

— the "speed" of travel as measured by arc length is constant

## ✍  Mastery Check 5.5:

Calculate the length of the 3D helical curve described by the vector function $\boldsymbol{r}(t) = \cos t\,\mathbf{i} + \sin t\,\mathbf{j} + t/\pi\,\mathbf{k}$, $0 \le t \le 2\pi$.

✍

## II.  Surfaces $\boldsymbol{r} : \mathbb{R}^2 \longrightarrow \mathbb{R}^3$

Another important class of vector-valued functions comprises those that depend on *two* independent variables, which we denote generically by $u$ and $v$. That is, we consider the class of vector-valued functions $\boldsymbol{f} : \mathbb{R}^2 \longrightarrow \mathbb{R}^m$, which map points $(u, v)$ in the plane to points in $m$-space:

$$(u, v) \longmapsto \boldsymbol{f}(u, v) = \big(f_1(u, v), f_2(u, v), ..., f_m(u, v)\big).$$

Since functions from $\mathbb{R}^2$ to $\mathbb{R}^3$ have great physical significance and are most readily illustrated, we limit the following discussion to these. As in **I**, we present the function as a point in $\mathbb{R}^3$, $\boldsymbol{r} = (x, y, z)$, all components of which depend on the two variables $(u, v)$:

$$\boldsymbol{r}(u, v) = \big(x(u, v), y(u, v), z(u, v)\big).$$

This vector function maps points in the $uv$-plane to points in 3-space and thereby traces out a *surface* shown here in Figure 5.8.

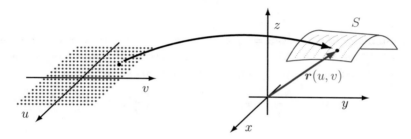

**Figure 5.8**  A surface in $\mathbb{R}^3$.

It is worth reinforcing here that in **I** we saw that a curve $\boldsymbol{r}(t)$ in 3D depends on *one* variable (Example 5.2). Here we see that a surface $\boldsymbol{r}(u, v)$ in 3D depends on *two* variables. The distinction is worth remembering as we shall have occasion to invoke a dimensional reduction (see the Remarks immediately following). Incidentally, the expression above, as well as most of the analysis to follow, supposes that $x$, $y$, $z$ are explicit functions of $u$ & $v$. However, in practice this may not always be the case; a surface may be defined implicitly (see the discussion on implicit functions in Section 2.H).

■  **Example 5.3:**

The set of parametric equations

$$\left.\begin{array}{l} x = a \cos u \sin v \\ y = a \sin u \sin v \\ z = a \cos v \end{array}\right\} \quad \text{satisfies} \quad x^2 + y^2 + z^2 = a^2.$$

This is a parametric representation of a sphere of radius $a$ centred at the origin.                                                                    ■

**Remarks**

* Before continuing, the reader might find it useful to revisit the discussions on coordinate systems and visualization of surfaces in Sections 1.D and 1.E.

* If we keep $u = u_0$ fixed we get $\boldsymbol{r} = \boldsymbol{r}(u_0, v)$, a vector function of one variable, $v$ (Figure 5.9). That is, restricting the variable $u$ results in a curve on $S$ called the constant $u$ curve. By the foregoing section this curve has a *tangent vector* given by

$$\boldsymbol{r}'_v(u_0, v) = \frac{\partial \boldsymbol{r}}{\partial v}(u_0, v) = \left( \frac{\partial x}{\partial v}(u_0, v), \frac{\partial y}{\partial v}(u_0, v), \frac{\partial z}{\partial v}(u_0, v) \right).$$

* Similarly, if we keep $v = v_0$ fixed we get $\boldsymbol{r} = \boldsymbol{r}(u, v_0)$, a vector function of the single variable $u$. This too is a curve on $S$, called the constant $v$ curve. Analogously, this curve has a tangent vector given by

$$\boldsymbol{r}'_u(u, v_0) = \frac{\partial \boldsymbol{r}}{\partial u}(u, v_0) = \left( \frac{\partial x}{\partial u}(u, v_0), \frac{\partial y}{\partial u}(u, v_0), \frac{\partial z}{\partial u}(u, v_0) \right).$$

* If $\boldsymbol{r}'_u(u_0, v_0) \times \boldsymbol{r}'_v(u_0, v_0) \neq 0$, which is the case for independent variables, then $\boldsymbol{r}'_u \times \boldsymbol{r}'_v$ is a vector normal to $S$ and normal to the *tangent plane* to $S$ (at the point $\boldsymbol{r}(u_0, v_0)$) spanned by the vectors $\boldsymbol{r}'_u(u_0, v_0)$ and $\boldsymbol{r}'_v(u_0, v_0)$.

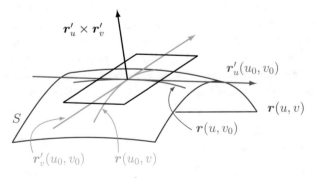

**Figure 5.9** The tangent plane to $S$ spanned by tangent vectors $\boldsymbol{r}'_u$ and $\boldsymbol{r}'_v$.

## III. The most general case $\boldsymbol{f} : \mathbb{R}^n \longrightarrow \mathbb{R}^m$

Although applications arise in more general cases of $\boldsymbol{f} : \mathbb{R}^n \longrightarrow \mathbb{R}^m$, we derive no benefit by specializing any further. We can instead reflect on the parallels that may be drawn between "projections" of a more general scenario and the special cases we have already discussed.

A differentiable vector function $\boldsymbol{f} : \mathbb{R}^n \longrightarrow \mathbb{R}^m$ with components $f_i$, $i = 1, 2, 3, \ldots, m$, which are real-valued functions of $\boldsymbol{x}$, is a vector-valued function

$$\boldsymbol{f}(\boldsymbol{x}) = \big(f_1(\boldsymbol{x}), f_2(\boldsymbol{x}), f_3(\boldsymbol{x}), \ldots, f_m(\boldsymbol{x})\big).$$

When we say that $\boldsymbol{x} \in \mathbb{R}^n$ is in the domain of $\boldsymbol{f}$, where $R_f \subset \mathbb{R}^m$, we mean that $\boldsymbol{x}$ is in the domain of each component, the scalar functions $f_i$, $i = 1, 2, \ldots, m$. Then if we can assume that each of the $f_i$ is continuous and has continuous partial derivatives, we can compute the gradient for each component,

$$\boldsymbol{\nabla} f_i(\boldsymbol{x}) = \left( \frac{\partial f_i}{\partial x_1}, \ldots, \frac{\partial f_i}{\partial x_n} \right), \ i = 1, 2, \ldots, m.$$

We have met the gradient in Section 2.E, where it was used to determine the rate of change of a scalar function in a specified direction. So, for instance, for the case $n = 2$ and $m = 1$, we had $z = f(x, y)$ describing a surface in 3D space, and the rate of change of $z$ at a point $(x_0, y_0)$ in the direction of unit vector $\boldsymbol{u} = (u, v)$ was given by

$$D_{\boldsymbol{u}} f(x_0, y_0) = \boldsymbol{\nabla} f\big|_0 \cdot \boldsymbol{u} = \left( \frac{\partial f}{\partial x}\Big|_0 \mathbf{i} + \frac{\partial f}{\partial y}\Big|_0 \mathbf{j} \right) \cdot (u\mathbf{i} + v\mathbf{j}),$$

or, in matrix notation,

$$D_{\boldsymbol{u}} f(x_0, y_0) = \left[ \frac{\partial f}{\partial x}\Big|_0 \ \frac{\partial f}{\partial y}\Big|_0 \right] \begin{pmatrix} u \\ v \end{pmatrix}.$$

With vector functions $\boldsymbol{f}(\boldsymbol{x})$ we have the potential to simultaneously find the rate of change of more that one scalar function using matrix multiplication. So for the case $n = 2$ and $m = 3$ we would have

$$D_{\boldsymbol{u}} \boldsymbol{f}(x_0, y_0) = \begin{bmatrix} \dfrac{\partial f_1}{\partial x}\Big|_0 & \dfrac{\partial f_1}{\partial y}\Big|_0 \\[2mm] \dfrac{\partial f_2}{\partial x}\Big|_0 & \dfrac{\partial f_2}{\partial y}\Big|_0 \\[2mm] \dfrac{\partial f_3}{\partial x}\Big|_0 & \dfrac{\partial f_3}{\partial y}\Big|_0 \end{bmatrix} \begin{pmatrix} u \\ v \end{pmatrix}.$$

The $3 \times 2$ matrix on the right is an example of a *Jacobian* matrix. The Jacobian lies at the heart of every generalization of single-variable calculus to higher dimensions.

**Definition 5.7**

Let $\boldsymbol{f} : \mathbb{R}^n \longrightarrow \mathbb{R}^m$ be a $C^1$ vector-valued function. The $m \times n$ matrix of first derivatives of $\boldsymbol{f}$ is called the **Jacobian matrix**.

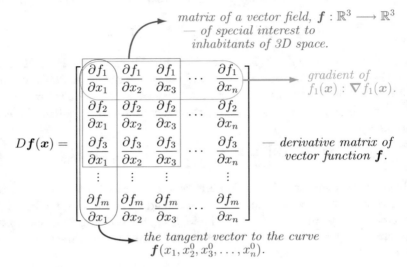

*matrix of a vector field, $\boldsymbol{f} : \mathbb{R}^3 \longrightarrow \mathbb{R}^3$*
*— of special interest to*
*inhabitants of 3D space.*

*gradient of*
*$f_1(\boldsymbol{x}) : \boldsymbol{\nabla} f_1(\boldsymbol{x})$.*

*— derivative matrix of*
*vector function $\boldsymbol{f}$.*

*the tangent vector to the curve*
*$\boldsymbol{f}(x_1, x_2^0, x_3^0, \ldots, x_n^0)$.*

**Figure 5.10** The Jacobian matrix of $\boldsymbol{f} : \mathbb{R}^n \longrightarrow \mathbb{R}^m$.

**Remarks**

* The $j^{\text{th}}$ row of the Jacobian in Figure 5.10 is the *gradient* of a scalar function $f_j(\boldsymbol{x})$, and there are $m$ of them.

* The $i^{\text{th}}$ column of the Jacobian in Figure 5.10 is a *tangent vector* to a curve in $\mathbb{R}^m$, and there are $n$ of them.

* The first three rows and first three columns correspond to the derivative matrix of a 3D vector field. (See Section 5.B.)

■ **Example 5.4:**

For a scalar function $f : \mathbb{R}^n \longrightarrow \mathbb{R}$, $\boldsymbol{x} \longmapsto f(\boldsymbol{x})$, the differential is

$$\mathrm{d}f = \frac{\partial f}{\partial x_1}\mathrm{d}x_1 + \frac{\partial f}{\partial x_2}\mathrm{d}x_2 + \cdots + \frac{\partial f}{\partial x_n}\mathrm{d}x_n = \left(\frac{\partial f}{\partial x_1}, \frac{\partial f}{\partial x_2}, \ldots, \frac{\partial f}{\partial x_n}\right)\begin{pmatrix} \mathrm{d}x_1 \\ \mathrm{d}x_2 \\ \vdots \\ \mathrm{d}x_n \end{pmatrix}.$$

In the case of a vector function $\boldsymbol{f} : \mathbb{R}^n \longrightarrow \mathbb{R}^m$, $\boldsymbol{x} \longmapsto \boldsymbol{f}(\boldsymbol{x})$, the differential generalizes to this:

$$
\mathrm{d}\boldsymbol{f} = \begin{pmatrix} \dfrac{\partial f_1}{\partial x_1} & \cdots & \dfrac{\partial f_1}{\partial x_n} \\ \vdots & & \vdots \\ \dfrac{\partial f_m}{\partial x_1} & \cdots & \dfrac{\partial f_m}{\partial x_n} \end{pmatrix} \begin{pmatrix} \mathrm{d}x_1 \\ \mathrm{d}x_2 \\ \vdots \\ \mathrm{d}x_n \end{pmatrix},
$$

— a matrix-vector product giving a vector.

■

■   **Example 5.5:**

For a scalar function $f : \mathbb{R}^n \longrightarrow \mathbb{R}$, $\boldsymbol{x} \longmapsto f(\boldsymbol{x})$, the chain rule gives

$$
\frac{\partial f}{\partial u_i} = \frac{\partial f}{\partial x_1}\frac{\mathrm{d}x_1}{\mathrm{d}u_i} + \frac{\partial f}{\partial x_2}\frac{\mathrm{d}x_2}{\mathrm{d}u_i} + \cdots + \frac{\partial f}{\partial x_n}\frac{\mathrm{d}x_n}{\mathrm{d}u_i} = \left(\frac{\partial f}{\partial x_1}, \frac{\partial f}{\partial x_2}, \ldots, \frac{\partial f}{\partial x_n}\right) \begin{pmatrix} \dfrac{\mathrm{d}x_1}{\mathrm{d}u_i} \\ \vdots \\ \dfrac{\mathrm{d}x_n}{\mathrm{d}u_i} \end{pmatrix}.
$$

In the case of a vector function $\boldsymbol{f} : \mathbb{R}^n \longrightarrow \mathbb{R}^m$, $\boldsymbol{x} \longmapsto \boldsymbol{f}(\boldsymbol{x})$, the chain rule generalizes to this:

$$
D(\boldsymbol{f} \circ \boldsymbol{x})(\boldsymbol{u}) = \begin{pmatrix} \dfrac{\partial f_1}{\partial x_1} & \cdots & \dfrac{\partial f_1}{\partial x_n} \\ \vdots & & \vdots \\ \dfrac{\partial f_m}{\partial x_1} & \cdots & \dfrac{\partial f_m}{\partial x_n} \end{pmatrix} \begin{pmatrix} \dfrac{\partial x_1}{\partial u_1} & \cdots & \dfrac{\partial x_1}{\partial u_k} \\ \vdots & & \vdots \\ \dfrac{\partial x_n}{\partial u_1} & \cdots & \dfrac{\partial x_n}{\partial u_k} \end{pmatrix}
$$

$$
= D\boldsymbol{f}(\boldsymbol{x}) \cdot D\boldsymbol{x}(\boldsymbol{u})
$$

— a matrix product giving a matrix.

■

# 5.B  Vector fields

A 3D vector-valued function of a 3D vector variable,

$$\boldsymbol{f} : D_f \subseteq \mathbb{R}^3 \longrightarrow \mathbb{R}^3, \ \boldsymbol{x} \longmapsto \boldsymbol{y} = \boldsymbol{f}(\boldsymbol{x}),$$

has special significance in physics and engineering, and other applications in the real world. Hence, it is given a special name: a *vector field*. To be explicit, an arbitrary vector field has the form

$$\boldsymbol{f}(\boldsymbol{x}) = \big(f_1(x, y, z), f_2(x, y, z), f_3(x, y, z)\big)$$

$$= f_1(x, y, z)\mathbf{i} + f_2(x, y, z)\mathbf{j} + f_3(x, y, z)\mathbf{k}$$

where the $f_1$, $f_2$, and $f_3$ are scalar functions of the three variables $x, y, z$.

Note that the subscripts "1", "2", and "3", do *not* here refer to partial derivatives, they refer to the components of our vector field.

Unless otherwise stated, we shall assume that the vector fields we work with have *continuous partial derivatives of order $m \geq 2$*. We will often refer to these as *smooth* and presume the component functions are $C^2$ or better.

**Some examples from physics**

(i) **Gravitational field**
    The gravitational force per unit mass ($G$ is the gravitational constant) is a 3D vector field (Figure 5.11).

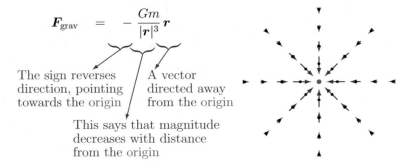

$$\boldsymbol{F}_{\text{grav}} \quad = \quad -\frac{Gm}{|\boldsymbol{r}|^3}\boldsymbol{r}$$

The sign reverses direction, pointing towards the origin

A vector directed away from the origin

This says that magnitude decreases with distance from the origin

**Figure 5.11** The gravitational field of a point mass.

## (ii) **Electrostatic fields**

(a) The 3D electrostatic field intensity (force per unit charge) is expressed in SI units by

$$\boldsymbol{E} = \frac{q}{4\pi\epsilon_0|\boldsymbol{r}|^3}\,\boldsymbol{r} \qquad \text{— due to a point charge } q \text{ at the origin.}$$

The direction depends on the sign of $q$. The constant $\epsilon_0$ is the "permittivity" of free space.

(b) The corresponding 2D version has the form,

$$\boldsymbol{E} = \frac{\rho}{2\pi\epsilon_0|\boldsymbol{r}|^2}\,\boldsymbol{r}.$$

The latter case can be thought of as a field in 3D free space due to a uniformly charged wire of infinite length (Figure 5.12), with the quantity $\rho$ being the charge per unit length of the wire.

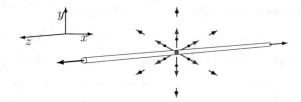

**Figure 5.12**  The electrostatic field near a charged wire.

## (iii) **Gradient field**

A vector field can also be derived from any spatially dependent scalar function by taking its gradient. For example, if $T = T(\boldsymbol{r})$ is a spatially varying temperature field, then $\boldsymbol{\nabla}T$ is the vector field, called the *temperature gradient*.

Gradient fields were discussed at some length in Section 2.E. They will recur often in this chapter, in both the differential and the integral contexts.

**Figure 5.13**  A field varying in direction and magnitude with position.

**Remarks**

* By now the reader will appreciate that both the magnitude and direction of a vector field will depend on position (Figure 5.13).

* As mentioned, Examples (i) and (ii)(a) are 3D vector fields $\mathbb{R}^3 \longrightarrow \mathbb{R}^3$.

* In Example (ii)(b), there are just *two* components to $\boldsymbol{E}$: $E_1 = E_1(x, y)$ and $E_2 = E_2(x, y)$. This is an example of a *plane vector field*

$$\boldsymbol{E} = E_1(x, y)\mathbf{i} + E_2(x, y)\mathbf{j}.$$

These will feature from time to time and specifically in Section 5.F.

* Example (iii) shows that one gets a vector field by taking the gradient of a real-valued scalar function, $\phi \in C^1(\mathbb{R}^3)$.

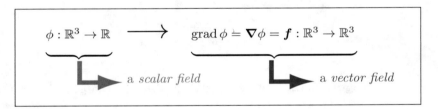

$$\phi : \mathbb{R}^3 \to \mathbb{R} \quad \longrightarrow \quad \operatorname{grad} \phi \stackrel{.}{=} \boldsymbol{\nabla} \phi = \boldsymbol{f} : \mathbb{R}^3 \to \mathbb{R}^3$$

*a scalar field*          *a vector field*

**More information about vector fields**

• **Divergence and curl**

The physical significance of the gradient was explained on Pages 76–78. However, there are two siblings of the gradient that are worth defining now in an operational sense as they too have physical meaning and mathematical utility. These are the *divergence* and *curl* of a vector field.

---

**Definition 5.8**
*Let $\boldsymbol{f} : \mathbb{R}^3 \longrightarrow \mathbb{R}^3$ be a $C^1$ (at least) vector field. Denote and define the* **divergence of $\boldsymbol{f}$** *by*

$$\operatorname{div} \boldsymbol{f} \ = \ \frac{\partial f_1}{\partial x} + \frac{\partial f_2}{\partial y} + \frac{\partial f_3}{\partial z} \equiv \boldsymbol{\nabla} \cdot \boldsymbol{f}.$$

---

The divergence thus operates on a vector field to give a *scalar* property of the field that applies locally. That is:

$$\underbrace{\boldsymbol{f} : \mathbb{R}^3 \to \mathbb{R}^3}_{} \quad\longrightarrow\quad \underbrace{\operatorname{div} \boldsymbol{f} \equiv \boldsymbol{\nabla} \cdot \boldsymbol{f} : \mathbb{R}^3 \to \mathbb{R}}_{}$$

a *vector field*            a *scalar field*

The notation employed in Definition 5.8 involving the gradient operator is standard and is used as a mnemonic to remind us of how the divergence is calculated, which is the expression shown in the central equality. The operator $\boldsymbol{\nabla}$ is treated as a vector in executing the scalar product with $\boldsymbol{F}$ although no actual component-wise multiplication is performed; instead, the components of $\boldsymbol{F}$ are partially differentiated with respect to the position variable corresponding to that component.

**Definition 5.9**

Let $\boldsymbol{f} : \mathbb{R}^3 \longrightarrow \mathbb{R}^3$ be a $C^1$ (at least) vector field. Denote and define the **curl of $\boldsymbol{f}$** by

$$\operatorname{curl} \boldsymbol{f} = \Big(\frac{\partial f_3}{\partial y} - \frac{\partial f_2}{\partial z}\Big)\mathbf{e}_1 + \Big(\frac{\partial f_1}{\partial z} - \frac{\partial f_3}{\partial x}\Big)\mathbf{e}_2 + \Big(\frac{\partial f_2}{\partial x} - \frac{\partial f_1}{\partial y}\Big)\mathbf{e}_3$$

$$= \begin{vmatrix} \mathbf{e}_1 & \mathbf{e}_2 & \mathbf{e}_3 \\ \dfrac{\partial}{\partial x} & \dfrac{\partial}{\partial y} & \dfrac{\partial}{\partial z} \\ f_1 & f_2 & f_3 \end{vmatrix} \equiv \boldsymbol{\nabla} \times \boldsymbol{f}.$$

In contrast to the divergence, the curl operates on a vector field to give a *vector* property of the field, also applied locally. That is:

$$\underbrace{\boldsymbol{f} : \mathbb{R}^3 \to \mathbb{R}^3}_{} \quad\longrightarrow\quad \underbrace{\operatorname{curl} \boldsymbol{f} \equiv \boldsymbol{\nabla} \times \boldsymbol{f} : \mathbb{R}^3 \to \mathbb{R}^3}_{}$$

a *vector field*            a *vector field*

As with the divergence operation, the notation employed in Definition 5.9 is standard form to help us remember how the curl is calculated, which is that shown on the right-hand side of the first equality. The operator $\boldsymbol{\nabla}$ is treated as a vector in the vector product with $\boldsymbol{F}$. Again, however, instead of pairwise multiplication, the components of $\boldsymbol{F}$ are partially differentiated with respect to the scalar variable corresponding to that location in the vector product.

Later we shall see that the *divergence* of a vector field is a measure of how much the field spreads out locally, while the *curl* of a vector field is a measure of how much the vector field turns or twists locally around an axis parallel to the direction of the vector $\boldsymbol{\nabla} \times \boldsymbol{F}$.

Note that in the case of two real vectors, switching the order of the vectors in the scalar product doesn't change the result of that product, while switching the orders of the vectors in the vector product results only in a vector pointing in the opposite direction. In the case of the divergence and curl, switching the order of $\boldsymbol{\nabla}$ and $\boldsymbol{F}$, *i.e.* writing $\boldsymbol{F} \cdot \boldsymbol{\nabla}$ and $\boldsymbol{F} \times \boldsymbol{\nabla}$ makes no sense if these appear in isolation. However, either expression may be legitimate if used in the context of an operation on another scalar field or another vector field, the latter in the case of another scalar or vector product, say. For example, the following makes perfect sense:

$$(\boldsymbol{F} \cdot \boldsymbol{\nabla})\phi = \left( F_1\frac{\partial}{\partial x} + F_2\frac{\partial}{\partial y} + F_3\frac{\partial}{\partial z} \right) \phi = F_1\frac{\partial \phi}{\partial x} + F_2\frac{\partial \phi}{\partial y} + F_3\frac{\partial \phi}{\partial x}$$

Analogous meanings can be ascribed to expressions such as $(\boldsymbol{F} \times \boldsymbol{\nabla}) \cdot \boldsymbol{G}$ and $(\boldsymbol{F} \times \boldsymbol{\nabla}) \times \boldsymbol{G}$. Remember, though, that the operations must be carried out in the correct order.

A large number of general and specific results of applications of the gradient, the divergence, the curl, and their combinations have been established. Some of these are listed on the next page. The reader is invited in Mastery Check 5.6 to prove some of these by direct application of the definitions.

- **Field lines**

  A useful concept particularly in fluid dynamics is that of field lines, also called *stream lines*. These are lines (actually *curves*) whose tangent vectors are parallel to the field vector at those points. That is, given that $\boldsymbol{r} : \mathbb{R} \longrightarrow \mathbb{R}^3, t \longmapsto \boldsymbol{r}(t)$ describes a curve in $\mathbb{R}^3$ (see page 224), then the field lines of a vector field $\boldsymbol{f}$ are defined by the equality:

$$\underbrace{\frac{\mathrm{d}\boldsymbol{r}}{\mathrm{d}t}}_{\text{tangent vector}} = \underbrace{\lambda(t)}\ \underbrace{\boldsymbol{f}\big(\boldsymbol{r}(t)\big)}_{\text{vector field}}$$

  tangent
  vector

  proportionality constant that is a
  function of position                                    (not usually of interest)

This vector equation hides three equations that must be solved simultaneously. If we assume the conditions of a smooth curve ($|\boldsymbol{r}'(t)| \neq 0$) then we deduce that $\lambda \neq 0$. Hence, we can solve the three component equations for $\lambda$ to get

$$\frac{1}{f_1(x,y,z)}\frac{\mathrm{d}x}{\mathrm{d}t} = \frac{1}{f_2(x,y,z)}\frac{\mathrm{d}y}{\mathrm{d}t} = \frac{1}{f_3(x,y,z)}\frac{\mathrm{d}z}{\mathrm{d}t}$$

$$\Rightarrow \frac{\mathrm{d}x}{f_1(x,y,z)} = \frac{\mathrm{d}y}{f_2(x,y,z)} = \frac{\mathrm{d}z}{f_3(x,y,z)}, \quad f_i \neq 0$$

Solving these three simultaneous differential equations, if possible, gives the curve described by $\boldsymbol{r}(t)$.

- **Some useful vector identities**

  Apart from the simple linear identities (*e.g.* $\boldsymbol{\nabla}\left(\phi + \psi\right) = \boldsymbol{\nabla}\phi + \boldsymbol{\nabla}\psi$), the operations of gradient, divergence, and curl obey a number of standard relations when applied to differentiable fields.

  Suppose vector fields $\boldsymbol{f}, \boldsymbol{g} : \mathbb{R}^3 \longrightarrow \mathbb{R}^3$ are $C^2$; vector $\boldsymbol{r} = (x, y, z)$; scalar functions $\phi, \psi : \mathbb{R}^3 \longrightarrow \mathbb{R}$ are $C^2$; scalar function $h : \mathbb{R} \longrightarrow \mathbb{R}$ is $C^1$; and $\boldsymbol{c} \in \mathbb{R}^3$ is a constant vector. Then the following identities can be readily derived:

1. $\boldsymbol{\nabla}\left(\phi\psi\right) = \psi\boldsymbol{\nabla}\phi + \phi\boldsymbol{\nabla}\psi$

2. $\boldsymbol{\nabla}\cdot\left(\phi\boldsymbol{f}\right) = \phi\boldsymbol{\nabla}\cdot\boldsymbol{f} + \boldsymbol{f}\cdot\boldsymbol{\nabla}\phi$

3. $\boldsymbol{\nabla}\times\left(\phi\boldsymbol{f}\right) = \phi\boldsymbol{\nabla}\times\boldsymbol{f} + \boldsymbol{\nabla}\phi\times\boldsymbol{f}$

4. $\boldsymbol{\nabla}\left(\boldsymbol{f}\cdot\boldsymbol{g}\right) = \left(\boldsymbol{f}\cdot\boldsymbol{\nabla}\right)\boldsymbol{g} + \left(\boldsymbol{g}\cdot\boldsymbol{\nabla}\right)\boldsymbol{f} + \boldsymbol{f}\times\left(\boldsymbol{\nabla}\times\boldsymbol{g}\right) + \boldsymbol{g}\times\left(\boldsymbol{\nabla}\times\boldsymbol{f}\right)$

5. $\boldsymbol{\nabla}\cdot\left(\boldsymbol{f}\times\boldsymbol{g}\right) = \boldsymbol{g}\cdot\left(\boldsymbol{\nabla}\times\boldsymbol{f}\right) - \boldsymbol{f}\cdot\left(\boldsymbol{\nabla}\times\boldsymbol{g}\right)$

6. $\boldsymbol{\nabla}\times\left(\boldsymbol{f}\times\boldsymbol{g}\right) = \boldsymbol{f}\left(\boldsymbol{\nabla}\cdot\boldsymbol{g}\right) - \boldsymbol{g}\left(\boldsymbol{\nabla}\cdot\boldsymbol{f}\right) + \left(\boldsymbol{g}\cdot\boldsymbol{\nabla}\right)\boldsymbol{f} - \left(\boldsymbol{f}\cdot\boldsymbol{\nabla}\right)\boldsymbol{g}$

7. $\boldsymbol{\nabla}\times\left(\boldsymbol{\nabla}\phi\right) = \boldsymbol{0}$

8. $\boldsymbol{\nabla}\cdot\left(\boldsymbol{\nabla}\times\boldsymbol{f}\right) = 0$

9. $\boldsymbol{\nabla}\times\left(\boldsymbol{\nabla}\times\boldsymbol{f}\right) = \boldsymbol{\nabla}\left(\boldsymbol{\nabla}\cdot\boldsymbol{f}\right) - \boldsymbol{\nabla}^2\boldsymbol{f}$

10. $\boldsymbol{\nabla}\cdot\boldsymbol{r} = 3$

11. $\boldsymbol{\nabla}h(r) = \dfrac{\mathrm{d}h}{\mathrm{d}r}\dfrac{\boldsymbol{r}}{r}$

12. $\boldsymbol{\nabla}\cdot\left(h(r)\boldsymbol{r}\right) = 3h(r) + r\dfrac{\mathrm{d}h}{\mathrm{d}r}$

13. $\boldsymbol{\nabla}\times\left(h(r)\boldsymbol{r}\right) = \boldsymbol{0}$

14. $\boldsymbol{\nabla}\left(\boldsymbol{c}\cdot\boldsymbol{r}\right)=\boldsymbol{c}$

15. $\boldsymbol{\nabla}\cdot\left(\boldsymbol{c}\times\boldsymbol{r}\right)=0$

16. $\boldsymbol{\nabla}\times\left(\boldsymbol{c}\times\boldsymbol{r}\right)=2\boldsymbol{c}$

### ✍ Mastery Check 5.6:
Confirm the vector identities 3, 5, 7, 8, 12 and 13.

✍

- **Conservative fields**

From the perspectives of physical significance and mathematical simplicity, one of the most important classes of vector fields is the class of so-called *conservative fields*.

---

**Definition 5.10**
*A vector field $\boldsymbol{f}:\ D\subseteq\mathbb{R}^3\longrightarrow\mathbb{R}^3$ is called a* **conservative vector field** *if there exists a $C^1$ scalar function $\phi:\ D\subseteq\mathbb{R}^3\longrightarrow\mathbb{R}$ such that*

$$\boldsymbol{f}(x,y,z)=\boldsymbol{\nabla}\phi(x,y,z).$$

*The function $\phi(x,y,z)$ is called a* **scalar potential** *of $\boldsymbol{f}$.*

---

Although scalar potentials and conservative fields arise in many areas of physics, it is far from true that all vector fields are conservative. That is, it is not generally true that all vector fields can be derived from scalar fields. In the next section we will discover an important and appealing mathematical property of conservative fields. For the moment we focus on the questions of establishing whether a vector field is conservative and if so what its scalar potential is.

To answer these questions we look at the properties of the scalar potential itself. Firstly, we see that for it to be a scalar potential, $\phi:D\longrightarrow\mathbb{R}$ must be at least $C^1$. Consequently, being a $C^1$ function, the differential of $\phi$ can be derived:

$$d\phi=\frac{\partial\phi}{\partial x}\,dx+\frac{\partial\phi}{\partial y}\,dy+\frac{\partial\phi}{\partial z}\,dz$$
$$=f_1\,dx+f_2\,dy+f_3\,dz.$$

The replacement of $\boldsymbol{\nabla}\phi$ with $\boldsymbol{f}$ in the last equation is valid since $\boldsymbol{f}=\boldsymbol{\nabla}\phi$ by assumption. The right-hand side of this last equation is thus an *exact differential* since it equals d$\phi$. That is, $f$ is conservative if $f_1 \, dx +$ $f_2 \, dy + f_3 \, dz$ is an exact differential. Moreover, if $\phi$ is a $C^2$ function, then (Definition 2.8, Page 83) we may conclude that

$$\frac{\partial^2 \phi}{\partial x \partial y} = \frac{\partial^2 \phi}{\partial y \partial x}, \quad \frac{\partial^2 \phi}{\partial x \partial z} = \frac{\partial^2 \phi}{\partial z \partial x}, \quad \frac{\partial^2 \phi}{\partial y \partial z} = \frac{\partial^2 \phi}{\partial z \partial y}.$$

Again, making the substitution $\nabla \phi = f$, we see that if $\phi$ is a potential to $f$, then the above equations are equivalent to:

$$\frac{\partial f_1}{\partial y} = \frac{\partial f_2}{\partial x}, \quad \frac{\partial f_3}{\partial x} = \frac{\partial f_1}{\partial z}, \quad \frac{\partial f_3}{\partial y} = \frac{\partial f_2}{\partial z}.$$

These are necessary conditions for $f = \nabla \phi$ to be true and thus for $f$ to be a conservative field. That is, the components of a conservative field must satisfy these interrelations. (See also Pages 258–260 and 288.)

✍  **Mastery Check 5.7:**

Determine whether the vector field $f = \left( xy - \sin z, \, \dfrac{x^2}{2} - \dfrac{e^y}{z}, \, \dfrac{e^y}{z^2} - x \cos z \right)$ is conservative, and if so, determine a potential to $f$.

Hints:

(1) For what values of $(x, y, z)$ is $f$ $C^1$?

(2) For those points $(x, y, z)$ confirm the necessary conditions for $f = \nabla \phi$.

(3) If possible, solve $\nabla \phi = f$ for a possible scalar potential $\phi$.

(4) Does $\phi$ exist? Everywhere? Does $\phi$ have any arbitrary constants? Are they unique?

✍

A scalar potential to a vector field $f$ is also a real-valued scalar function, and we have seen that (see Section 2.E) the *level surfaces* of $\phi : \phi(x, y, z) = c$ have normal vectors given by $\nabla \phi$. This means that for a conservative field, $f = \nabla \phi$ is a vector normal to the surface $\phi = c$ at $r$. The level surfaces of $\phi$ (defined in Section 1.F) are called *equipotential* surfaces of $f$. See Figure 5.14.

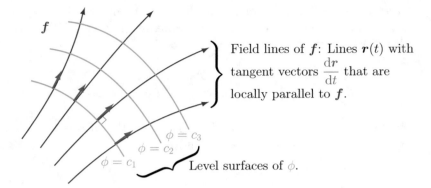

Field lines of $\boldsymbol{f}$: Lines $\boldsymbol{r}(t)$ with tangent vectors $\dfrac{\mathrm{d}\boldsymbol{r}}{\mathrm{d}t}$ that are locally parallel to $\boldsymbol{f}$.

$\phi = c_3$

$\phi = c_2$

$\phi = c_1$

Level surfaces of $\phi$.

**Figure 5.14** Conservative field lines and level surfaces of a scalar potential.

We leave differential vector calculus for the moment and consider some elements of the integral vector calculus, specifically so-called *curve* and *surface* integrals.

# 5.C   Line integrals

Line integrals are somewhat more complicated versions of one-dimensional integrals. As we alluded to in Chapter 4, the most obvious generalization involves replacing the 1D *interval* over which an integral is evaluated, with a one-variable parameterized curve in 3D. As a consequence our intuitive view of a 1D integral as an area under a curve is no longer applicable. We need to rely on a visual idea. Given their respective physical applications, line integrals can be divided into two classes. Within each class it is possible to utilize a specific physical picture to help engender an appreciation for that class of integral.

## I. Line integrals of real-valued functions

### Physical motivation

In its simplest description these are concerned with one-dimensional integrals of scalar functions to evaluate the total measure of something that is distributed along a curve.

Recall from single-variable calculus that

$$\int_a^b f(x)\, dx$$

is motivated as giving the area "under" $f$ and "over" straight-line interval $[a, b]$

— this is a geometric interpretation

Suppose instead that we interpret $f(x)$ as a variable linear density of some quantity (e.g. mass/unit length or charge/unit length) defined along the interval $I = [a, b]$. Then

$$\int_I f(x)\, dx$$

would give the total amount of that quantity, say total mass, attributed to $I$

— this is a physical interpretation

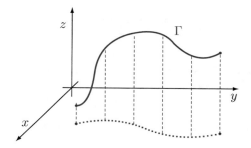

**Figure 5.15**  The curve $\Gamma$ in 3D and its projection onto the 2D plane.

Now let's extend the latter idea to higher dimensions. Suppose we were interested in evaluating the total mass of a nonlinear, one-dimensional object in 3D, one that is represented by the curve, $\Gamma$, in $\mathbb{R}^3$ (Figure 5.15).

What we want is an expression for the total mass of the curved object which corresponds to $\int_I f(x)\, dx$ for a straight-line interval.

**Mathematical construction**

Suppose $\Gamma$ is a finite *smooth* curve in $\mathbb{R}^3$. Then, from our discussion in Section 5.A.I, there exists a (non-unique) one-to-one parametrization of $\Gamma$,

$$\boldsymbol{r}(t) = \big(x(t), y(t), z(t)\big), \quad t \in [a, b],$$

such that $\boldsymbol{r}(t)$ is continuously defined on the finite connected interval $[a, b]$, and $|\boldsymbol{r}'(t)| \neq 0$.

Consider a partition of $[a,b]$, $a = t_0 < t_1 < t_2 < \cdots < t_{n-1} < t_n = b$, which leads to a set of discrete points on $\Gamma$: $\{\boldsymbol{r}(t_0), \boldsymbol{r}(t_1), \ldots, \boldsymbol{r}(t_n)\}$, as shown in Figure 5.16.

**Figure 5.16**  A parametrization of $\Gamma$.

The leading-order distance between nearest neighbour (in $t$) points on $\Gamma$ is $|\Delta \boldsymbol{r}_i| = |\boldsymbol{r}(t_i) - \boldsymbol{r}(t_{i-1})|$ and adding all $n$ such line segment contributions gives

$$\sigma_n = \sum_{i=1}^{n} |\Delta \boldsymbol{r}_i| = \sum_{i=1}^{n} |\boldsymbol{r}(t_i) - \boldsymbol{r}(t_{i-1})|$$

as an approximation to the total length of $\Gamma$. Since the curve is smooth w.r.t. $t$, we can apply the mean value theorem to each component function of $\boldsymbol{r}$ to get

$$|\Delta \boldsymbol{r}_i| = |\boldsymbol{r}(t_i) - \boldsymbol{r}(t_{i-1})| = |(x'(\zeta_i), y'(\eta_i), z'(\xi_i))| \Delta t_i$$

where $\{\zeta_i, \eta_i, \xi_i\}$ are some values of $t$, not necessarily the same, in the interval $(t_{i-1}, t_i)$. We then have

$$\sigma_n = \sum_{i=1}^{n} |(x'(\zeta_i), y'(\eta_i), z'(\xi_i))| \Delta t_i.$$

Now taking the dual limit of an infinite number of partition intervals and of vanishing partition size we get the total length of $\Gamma$:

$$|\Gamma| = \lim_{\substack{n \to \infty \\ \max|\Delta t_i| \to 0}} \sum_{i=1}^{n} |\Delta \boldsymbol{r}_i| = \int_a^b |\boldsymbol{r}'(t)| \, \mathrm{d}t$$

— provided the limit exists, which it should. (Why?)

We now extend the above argument to include a curve position-dependent function.

Let $f : \mathbb{R}^3 \longrightarrow \mathbb{R}$ be a continuous scalar function defined on some domain in $\mathbb{R}^3$. When restricted to $\Gamma$ in that domain it becomes a composite function of one variable:

$$f(x(t), y(t), z(t)) = f(\boldsymbol{r}(t)).$$

With this restriction $f$ can be thought of as a local linear density of some property of $\Gamma$ (again, mass or charge per unit length). Multiplying $f$ with a segment of length $|\Delta r|$ then gives the total amount of that property (mass or charge) possessed by that segment.

By an argument analogous to the above, but applied then to $f(r)|\Delta r|$, we find that the integral measure of the property represented by $f$ is given by ($t_i^* \in [t_{i-1}, t_i]$)

$$\lim_{\substack{n \to \infty \\ \max(|\Delta r|) \to 0}} \sum_{i=1}^{n} f(x(t_i^*), y(t_i^*), z(t_i^*))|\Delta r_i| = \int_a^b f(x(t), y(t), z(t))|r'(t)| \, dt$$

— provided the limit exists, which it
will if $f$ is continuous and $[a, b]$
is bounded (Theorem 3.2).

## Remarks

* It should be obvious from our derivation of curve length and the extension to the total mass that integrals along a given curve of *scalar* functions have *no dependence* on the *direction* taken along the curve. We could have started at either end and arrived at the same result provided the parametrization is defined in a one-to-one manner, increasing from the start position to the end position. This is true of all curve integrals of scalar functions.

* If we had considered a sub-interval $[a, \tau] \subset [a, b]$, the above argument would have led to Definition 5.6 (Page 231) for the *arc length* $s(\tau)$:

$$s(\tau) = \int_a^\tau |r'(t)| \, dt.$$

with $s(a) = 0$ and $s(b) = |\Gamma|$. As shown on Page 231, an application of Leibniz's rule for the derivative of an integral gives

$$\frac{ds}{d\tau} = \left| \frac{dr}{d\tau} \right|.$$

With a convenient renaming of independent variable $\tau \longrightarrow t$ we are led to the differential arc length

$$ds = |dr| = \left| \frac{dr}{dt} \right| dt$$

Consequently, in terms of this differential arc length (equivalent to the parametric integral derived above), the total length of $\Gamma$ is

$$|\Gamma| = \int_\Gamma \mathrm{d}s \quad \left(= s(b) = \int_a^b \left|\frac{\mathrm{d}\boldsymbol{r}}{\mathrm{d}t}\right| \mathrm{d}t\right),$$

while the integral of $f(\boldsymbol{r})$ over $\Gamma$ is

$$\int_\Gamma f(\boldsymbol{r})\mathrm{d}s \quad \left(= \int_a^b f\big(\boldsymbol{r}(t)\big)\left|\frac{\mathrm{d}\boldsymbol{r}}{\mathrm{d}t}\right| \mathrm{d}t\right).$$

∗ Neither integral can depend on how we define $\Gamma$, *i.e.* they cannot depend on what parametrization we choose.

### ■  Example 5.6:

Determine the integral $\displaystyle\int_\Gamma (xy + y)\,\mathrm{d}s$ where $\Gamma$ is the path along the 2D curve $y = \sqrt{x}$ from the point $(4, 2)$ to the point $(9, 3)$.

Let $\boldsymbol{r}(x, y)$ define a point on $\Gamma$. We parameterize $\boldsymbol{r} = (x, y)$ as $(t, \sqrt{t})$, with $t : 4 \to 9$, and $\dfrac{\mathrm{d}\boldsymbol{r}}{\mathrm{d}t} = \left(1, \dfrac{1}{2\sqrt{t}}\right)$.

Then $\left|\dfrac{\mathrm{d}\boldsymbol{r}}{\mathrm{d}t}\right| = \sqrt{1 + \dfrac{1}{4t}}$ and

$$\begin{aligned}
\int_\Gamma (xy + y)\,\mathrm{d}s &= \int_4^9 \left(t^{3/2} + t^{1/2}\right)\sqrt{1 + \frac{1}{4t}}\,\mathrm{d}t \;=\; \frac{1}{2}\int_4^9 (t + 1)\sqrt{4t + 1}\,\mathrm{d}t \\
&= \left[(t + 1)\frac{1}{6}(4t + 1)^{3/2}\right]_4^9 - \int_4^9 \frac{1}{6}(4t + 1)^{3/2}\,\mathrm{d}t \\
&= \left[(t + 1)\frac{1}{6}(4t + 1)^{3/2} - \frac{1}{60}(4t + 1)^{5/2}\right]_4^9 \\
&= \left[\frac{1}{20}(2t + 3)(4t + 1)^{3/2}\right]_4^9 \;=\; \frac{1}{20}\left(777\sqrt{37} - 187\sqrt{17}\right).
\end{aligned}$$

■

### ✍  Mastery Check 5.8:

Evaluate $\displaystyle\int_C x^2 y^2\,\mathrm{d}s$ where $C$ is the full circle $x^2 + y^2 = 2$.

Hint: A suitable parametrization is with polar coordinates. In this case $\boldsymbol{r}(t) = \sqrt{2}\cos t\,\boldsymbol{e}_1 + \sqrt{2}\sin t\,\boldsymbol{e}_2$, $t : 0 \to 2\pi$.

✍

## II. Line integrals of vector-valued functions

In this class of line integrals we not only deal with vector fields as opposed to scalar fields, but we take into consideration the relation between the direction of the integration path taken and the field direction.

### Physical motivation

Imagine a block of mass $m$ sitting on a table under gravity. Suppose a horizontal force of magnitude $F$ is needed to slide the block against friction (Figure 5.17(a)).

To move the block a distance $d$ under a constant force $F$ and in the same direction as $F$, the work done is $W = F.d$

If there is no friction, the only external force is gravity and no work is done in sliding the block horizontally.

However, in moving the block directly upwards against gravity (Figure 5.17(b)), the work done is proportional to the force required to overcome gravity. The work done is equal to the force required to overcome gravity ($mg$) times distance travelled ($h$): $W = mgh$.

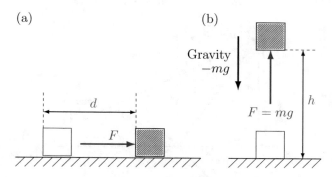

**Figure 5.17**   (a) Horizontal displacement opposing friction;
(b) Vertical displacement opposing gravity.

If we want to move the block a distance $d$ in a straight line in an arbitrary direction when the only force is gravity, we would then have the scenario pictured in Figure 5.18. The work done depends *only* on the vertical component of the displacement: $W = mgh = mgd\sin\phi$.

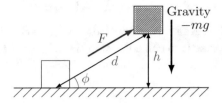

**Figure 5.18** Displacement in an oblique direction.

In 3D space coordinates, where the block moves from $A$ to $B$, and the only opposing force is gravity, the situation can be imagined as in Figure 5.19. The work done is now

$$W = mg.\Delta h = mg.d.\sin\phi$$
$$= mg.d.\sin\left(\frac{\pi}{2} - \theta\right)$$
$$= mg.d.\cos\theta$$
$$= \mathbf{F}_g \cdot (\mathbf{r}_B - \mathbf{r}_A).$$

That is, it is simply but significantly the scalar product of the vector force needed to overcome gravity, and the vector displacement.

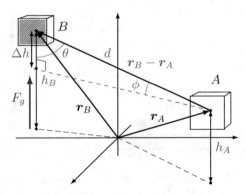

**Figure 5.19** Displacement in 3D space.

Generally, the amount of work done depends on the direction of the displacement and the direction of the applied force.

If $F \perp$ displacement, then $W = 0$

If $F \parallel$ (+)displacement, then $W$ is the maximum $= F.d$

As we have seen, for a more general direction of displacement we have the intermediate case:

$$W = \mathbf{F} \cdot \mathbf{d} = Fd\cos\theta.$$

(a)                                              (b)

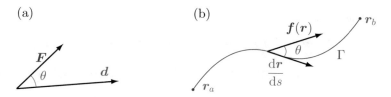

**Figure 5.20**  (a) Constant force and displacement;
(b) Incremental displacement along a curve.

So far, we have assumed the force and displacement are constant in terms of both direction and magnitude, Figure 5.20(a). But what happens when the force $f$ is not constant but a function of position, $f = f(r)$, and the displacement is along a more variable path $\Gamma$?

**Mathematical construction**

Suppose $r = r(s)$ describes the path to be taken (Figure 5.20(b)). Consider a small segment of the curve, specifically the arc length differential $ds$ at $r(s)$.

Over the *infinitesimal* curve segment, $ds$, with unit *tangent vector*, $\dfrac{dr}{ds}$, $f$ is approximately constant, and $ds$ is approximately straight. Then the work increment to leading order in $ds$ will be

$$dW = |f| \cos\theta \, ds$$
$$= f(r) \cdot \frac{dr}{ds} ds$$
$$= f(r) \cdot dr.$$

This means that the total work done in moving from start to finish along $\Gamma$ will be the integral

$$W = \int_\Gamma dW = \int_\Gamma \left( f(r) \cdot \frac{dr}{ds} \right) ds = \int_\Gamma f(r) \cdot dr.$$

**Remarks**

* $\displaystyle\int_\Gamma f \cdot dr$ is called the line integral of the vector field $f$ (or its tangential component) along $\Gamma$.

* Since $\left| \dfrac{dr}{ds} \right| = 1$ (see Page 232), then $f(r) \cdot \dfrac{dr}{ds} = |f| \cos\theta$.

— the tangential component of $f$ along $\Gamma$

* In Cartesian form

$$\int_\Gamma \boldsymbol{f} \cdot \mathrm{d}\boldsymbol{r} = \int_\Gamma \Big( f_1(x,y,z)\mathrm{d}x + f_2(x,y,z)\mathrm{d}y + f_3(x,y,z)\mathrm{d}z \Big).$$

— this is needed if the path is specified in $x, y$ and $z$

* For an arbitrary parametrization, $\boldsymbol{r}(t)$, with $t \in [a,b]$, $\boldsymbol{r}_a = \boldsymbol{r}(a)$, and $\boldsymbol{r}_b = \boldsymbol{r}(b)$, we have (for a smooth curve) the very practical expression

$$\int_\Gamma \boldsymbol{f} \cdot \mathrm{d}\boldsymbol{r} = \int_a^b \boldsymbol{f}\big(\boldsymbol{r}(t)\big) \cdot \frac{\mathrm{d}\boldsymbol{r}(t)}{\mathrm{d}t}\mathrm{d}t.$$

as a one-dimensional integral!

* Work is positive or negative depending on the direction of motion; the sign dictating whether energy must be applied to achieve the displacement against the force (negative) or is gained in moving with the force (positive). Therefore, in contrast to line integrals of scalar functions, for line integrals of vector functions we *must always* specify the direction of displacement.

Since $\boldsymbol{f} \cdot (-\mathrm{d}\boldsymbol{r}) = -\boldsymbol{f} \cdot \mathrm{d}\boldsymbol{r}$, we find that

$$\int_\Gamma \boldsymbol{f} \cdot \mathrm{d}\boldsymbol{r} = -\int_{-\Gamma} \boldsymbol{f} \cdot \mathrm{d}\boldsymbol{r}$$

A curve with a specified direction is called *oriented* (Figure 5.21).

– this is important to remember in calculations.

**Figure 5.21** Displacement directions along a curve.

* $\int_\Gamma \boldsymbol{f} \cdot \mathrm{d}\boldsymbol{r}$ is independent of the choice of parametrization — as long as we go in the *same* direction.

* If $\Gamma = \Gamma_1 \cup \Gamma_2 \cup \Gamma_3 \cup \cdots \cup \Gamma_n$, as in Figure 5.22, then

$$\int_\Gamma \boldsymbol{f} \cdot \mathrm{d}\boldsymbol{r} = \int_{\Gamma_1} \boldsymbol{f} \cdot \mathrm{d}\boldsymbol{r} + \int_{\Gamma_2} \boldsymbol{f} \cdot \mathrm{d}\boldsymbol{r} + \int_{\Gamma_3} \boldsymbol{f} \cdot \mathrm{d}\boldsymbol{r} + \cdots + \int_{\Gamma_n} \boldsymbol{f} \cdot \mathrm{d}\boldsymbol{r}.$$

**Figure 5.22**  A connected, piecewise smooth, curve.

The total line integral over $\Gamma$ is the sum of the line integrals over the *connected* pieces $\Gamma_i$ that link the start (of $\Gamma$) to the end (of $\Gamma$).

* If $\Gamma$ is a closed curve (Figure 5.23), that is if $\boldsymbol{r}_a = \boldsymbol{r}_b$, then the line integral of $\boldsymbol{f}$ along $\Gamma$ is called the *circulation* of $\boldsymbol{f}$ around $\Gamma$:

Circulation of $\boldsymbol{f}$ around $\Gamma$ is $\displaystyle\oint_\Gamma \boldsymbol{f} \cdot \mathrm{d}\boldsymbol{r}$.

**Figure 5.23**  The circle: a simple closed curve.

In the next three examples we demonstrate the different ways of evaluating line integrals of vector fields.

■  **Example 5.7:**

Evaluating line integrals using Cartesian coordinates $(x, y, z)$.

We wish to evaluate $\displaystyle\int_\Gamma \boldsymbol{f} \cdot \mathrm{d}\boldsymbol{r}$ where $\boldsymbol{f} = (x + y, y - x)$ and $\Gamma$ is a curve with endpoints $(1, 1, 0)$ and $(4, 2, 0)$. We will choose two paths:

(a) $\Gamma = \{(x, y, z) : x = y^2, z = 0\}$ (Figure 5.24(a)), and
(b) $\Gamma = \Gamma_1 \cup \Gamma_2$, where $\Gamma_1$ and $\Gamma_2$ are paths of constant $y$ and constant $x$ as shown in Figure 5.24(b).

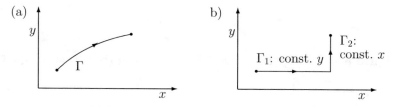

**Figure 5.24**  The paths (a) $x = y^2, z = 0$, and (b) $\Gamma = \Gamma_1 \cup \Gamma_2$.

a) Let $y = t$, $t \in [1, 2]$. Then $x = t^2$, $\boldsymbol{f} = (f_1, f_2, f_3) = (t^2 + t, t - t^2, 0)$,

$r = (t^2, t, 0)$, and $\dfrac{\mathrm{d}r}{\mathrm{d}t} = (2t, 1, 0)$.

Hence $\displaystyle\int_\Gamma f \cdot \mathrm{d}r = \int_1^2 (t^2 + t, t - t^2, 0) \cdot (2t, 1, 0)\,\mathrm{d}t = \dfrac{34}{3}$.

b) On $\Gamma_1$ let $x = t$, $t \in [1, 4]$. Then $y = 1$, $f = (f_1, f_2, f_3) = (t + 1,$ $1 - t, 0)$, $r = (t, 1, 0)$, and $\dfrac{\mathrm{d}r}{\mathrm{d}t} = (1, 0, 0)$. On $\Gamma_2$ let $y = t$, $t \in [1, 2]$. Then $x = 4$, $f = (f_1, f_2, f_3) = (4 + t, t - 4, 0)$, $r = (4, t, 0)$, and $\dfrac{\mathrm{d}r}{\mathrm{d}t} = (0, 1, 0)$.

Thus $\displaystyle\int_\Gamma f \cdot \mathrm{d}r = \int_{\Gamma_1} f \cdot \mathrm{d}r + \int_{\Gamma_2} f \cdot \mathrm{d}r$

$\displaystyle = \int_1^4 (t + 1, 1 - t, 0) \cdot (1, 0, 0)\,\mathrm{d}t + \int_1^2 (4 + t, t - 4, 0) \cdot (0, 1, 0)\,\mathrm{d}t$

$\displaystyle = \int_1^4 (t + 1)\,\mathrm{d}t + \int_1^2 (t - 4)\,\mathrm{d}t = 8.$

■

## ■ Example 5.8:

Evaluating line integrals using curve the parametrization $(x(t), y(t), z(t))$.

Evaluate $\displaystyle\int_\Gamma f \cdot \mathrm{d}r = \int_\Gamma (xz\,\mathrm{d}x + y^2\,\mathrm{d}y + x^2\,\mathrm{d}z)$ from $(0, 0, 0)$ to $(1, 1, 1)$.

along the curves (a) $r(t) = (t, t, t^2)$, and (b) $r(t) = (t^2, t, t^2)$ (Figure 5.25).

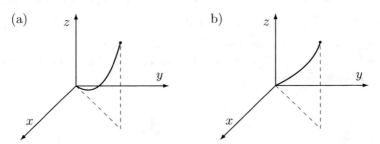

**Figure 5.25** The paths (a) $r(t) = (t, t, t^2)$, and (b) $r(t) = (t^2, t, t^2)$.

In both cases $t \in [0, 1]$ will work.

a) We have $x = t$, $y = t$, $z = t^2$, so $f = (t^3, t^2, t^2)$, $\dfrac{\mathrm{d}r}{\mathrm{d}t} = (1, 1, 2t)$, and

$\displaystyle\int_C f \cdot \mathrm{d}r = \int_0^1 (t^3, t^2, t^2) \cdot (1, 1, 2t)\,\mathrm{d}t = \int_0^1 (t^3 + t^2 + 2t^3)\,\mathrm{d}t = \dfrac{13}{12}.$

b) We have $x = t^2$, $y = t$, $z = t^2$, so $\boldsymbol{f} = (t^4, t^2, t^4)$, $\dfrac{\mathrm{d}\boldsymbol{r}}{\mathrm{d}t} = (2t, 1, 2t)$, and

$$\int_C \boldsymbol{f} \cdot \mathrm{d}\boldsymbol{r} = \int_0^1 (t^4, t^2, t^4) \cdot (2t, 1, 2t)\, \mathrm{d}t = \int_0^1 (2t^5 + t^2 + 2t^5)\, \mathrm{d}t = 1.$$

■

And here is the third example — for the reader to try.

✎  **Mastery Check 5.9:**

Evaluating curve integrals along curves defined by intersections of surfaces.
Evaluate $\displaystyle\oint_\Gamma \boldsymbol{f} \cdot \mathrm{d}\boldsymbol{r}$ where $\boldsymbol{f} = y\mathbf{i} + \mathbf{k}$ and $\Gamma$ is the curve of intersection of the cone

$$z^2 = 2x^2 + 2y^2$$

and the plane $y = z + 1$. See Figure 5.26.

Orientation is counterclockwise seen from $(0, 0, 117)$.  ✎

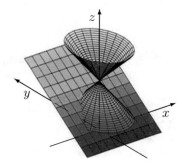

**Figure 5.26** Intersection of a plane and a cone.

✎

✎  **Mastery Check 5.10:**

Calculate the work done by a force $\boldsymbol{f}(\boldsymbol{r}) = y\,\mathbf{e}_1 + 2x\,\mathbf{e}_2 - z\,\mathbf{e}_3$ along a helical curve $\Gamma$ defined by $\boldsymbol{r}(t) = \cos t\,\mathbf{e}_1 + \sin t\,\mathbf{e}_2 + t\,\mathbf{e}_3$, $t : 0 \to 2\pi$.

✎

## Remarks

&ast; Each of the line integrals in Examples 5.7 and 5.8 and Mastery Check
5.9 eventually involve some form of parametrization of the curve, with
the result that the line integrals reduce to single integrals of functions
of one variable, the parameter.

&ast; In Examples 5.7 and 5.8 the line integrals are of the same field along
*different* curves joining the same endpoints. They give *different* results!

To help understand why this may happen it helps to defer to our phys-
ical motivation. The work done in going from $\boldsymbol{r}_a$ to $\boldsymbol{r}_b$ depends on the
path taken; different values result from different routes. Here we can
think of the action of *friction*.

The last remark raises an important point and an important question. What
conditions do we need to impose on a vector field to get the same result for
the line integral?

The satisfactory answer to this question is that the field must be *conservative*!

---

**Theorem 5.2**
*Suppose $\boldsymbol{f}(\boldsymbol{x}) = \big(f_1(\boldsymbol{x}), f_2(\boldsymbol{x}), f_3(\boldsymbol{x})\big)$ is a conservative field with potential
$\phi(\boldsymbol{x})$ defined on an open connected domain, $D$. Then*

$$\int_\Gamma \boldsymbol{f} \cdot \mathrm{d}\boldsymbol{r} = \phi(\boldsymbol{r}_b) - \phi(\boldsymbol{r}_a)$$

*for **every** curve $\Gamma$ lying entirely in $D$ which joins the points $\boldsymbol{r}_a \longrightarrow \boldsymbol{r}_b$.*

---

Theorem 5.2 states that the line integral $\int_\Gamma \boldsymbol{f} \cdot \mathrm{d}\boldsymbol{r}$ is independent of the choice
of $\Gamma$ if $\boldsymbol{f}$ *is conservative* and is dependent only on the endpoints, $\boldsymbol{r}_a$ and $\boldsymbol{r}_b$. To
be precise, the integral is determined by, and dependent only on, the values
of the potential $\phi$ at these points.

A proof of the theorem, and our consequential statements, follows readily
from our earlier finding that for a conservative field $\boldsymbol{f} \cdot \mathrm{d}\boldsymbol{r}$ is a perfect differential, $\mathrm{d}\phi$. The integral of this perfect differential results in the difference in
the values of the potential at the terminal points of the curve $\Gamma$.

**Figure 5.27**   Two curves joining the same end points.

**Corollary 5.2.1** *For the conditions of Theorem 5.2, let $\Gamma_1$ and $\Gamma_2$ be two curves in $D$ that join $r_a$ and $r_b$ (Figure 5.27). Then*

$$\int_{\Gamma_1} f \cdot dr = \int_{\Gamma_2} f \cdot dr$$

$$\implies \int_{\Gamma_1} f \cdot dr - \int_{\Gamma_2} f \cdot dr = 0$$

$$\implies \int_{\Gamma_1} f \cdot dr + \int_{-\Gamma_2} f \cdot dr = 0$$

$$\implies \int_{\Gamma_1 \cup (-\Gamma_2)} f \cdot dr = \oint_{\Gamma} f \cdot dr = 0.$$

Thus, the circulation of conservative field $f$ is zero:   $\oint_{\Gamma} f \cdot dr = 0.$

In the statement of Theorem 5.2 we impose the condition that $D$ must be a *connected domain*. This means that any pair of points $r_a$ and $r_b$ can be joined by a piecewise smooth curve which lies inside $D$.

Two examples of this, and a counterexample, are shown in Figure 5.28.

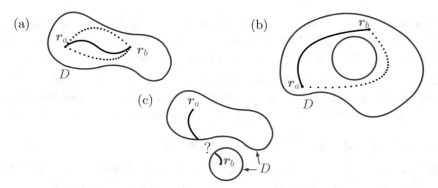

**Figure 5.28**   Connected domains (a) and (b).
(c) is a disjoint or disconnected domain.

**Remark**

&ast;   Theorem 5.2 and Corollary 5.2.1 are useful even for nonconservative fields — especially for *nearly* conservative fields. (See the next Mastery Check.)

✍  **Mastery Check 5.11:**

Let $\boldsymbol{f} = \left( Ax \ln z, By^2 z, \dfrac{x^2}{z} + y^3 \right)$.

(a) Determine $A$ and $B$ so that $\boldsymbol{f}$ is conservative.

(b) Evaluate $\displaystyle\int_{\Gamma} \left( 2x \ln z \, \mathrm{d}x + 2y^2 z \, \mathrm{d}y + y^3 \, \mathrm{d}z \right)$

   where $\Gamma$ is the straight line $(1,1,1) \to (2,1,2)$.

                                                                                          ✍

✍  **Mastery Check 5.12:**

Evaluate the line integral of $\boldsymbol{f}(\boldsymbol{r}) = x\mathbf{e}_1 + e^y \mathbf{e}_2$ along a curve $\Gamma$ where $\Gamma$ is
(a) a circular arc from $(0,2)$ to $(2,0)$, and
(b) a straight line connecting $(0,2)$ to $(2,0)$.
Is $\boldsymbol{f}$ conservative?

                                                                                          ✍

# 5.D   Surface integrals

In complete analogy with line integrals, surface integrals are generalized versions of double integrals! As with line integrals, surface integrals fall into one of two classes dictated by the function being integrated.

Instead of a double integral with an integration over a planar domain, with its interpretation as the volume under the graph of a function, a surface integral is effectively a sum of some quantity that is distributed over the surface itself.

## I. Surface integrals of real-valued functions

In this class the surface integrals operate on scalar functions and give the total amount of something that is distributed over a surface $S$ in $\mathbb{R}^3$.

Corollary 4.1.1 (2D version) for $f \geq 0$ gives the geometric interpretation of a double integral of $f$ as

$$\iint_D f(x,y)\, dA \;=\; \text{volume of body under } f \text{ over } D \text{ (a planar subset of } \mathbb{R}^2\text{).}$$

If, however, $f(x,y)$ were interpreted as, say, a charge/unit area (surface charge density) defined on $\mathbb{R}^2$ then the integral would give the total charge on the planar region $D$ in the $xy$-plane: $\iint_D f(x,y)\, dA \;=\; Q$ (Figure 5.29).

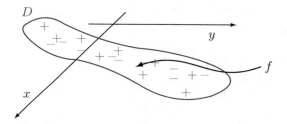

**Figure 5.29**  Physical interpretation of a double integral
over a planar region $D$.

Now suppose that instead of a region in the plane we have a surface $S \subset \mathbb{R}^3$ defined as a set of points

$$S = \{r(u,v) \in \mathbb{R}^3 : (u,v) \in D \subset \mathbb{R}^2\}.$$

For the present, we assume there is a one-to-one correspondence between points in $D$ and points on $S$ (this can be altered to consider piecewise mappings of parts of $S$ to subdomains if a one-to-one mapping of the whole of $S$ is not possible).

Moreover, suppose we are given a continuous scalar function $f : \mathbb{R}^3 \longrightarrow \mathbb{R}$ defined on a sub-domain in $\mathbb{R}^3$. Restricting $f$ to $S$ within that domain, the function becomes a composite function of two variables:

$$f(x(u,v), y(u,v), z(u,v)) = f(r(u,v)).$$

The function $f(r(u,v)$ applied to a bounded surface $S$ can be thought of as a summable surface area density (say, charge/unit area or mass/unit area), this time defined on $S$ rather than $D$. Multiplying $f$ with a segment of area $\Delta S$ gives the total amount of charge possessed by that segment.

Similarly to how we reasoned with line integrals, we partition the surface $S$ into a number $n$ of small segments, $\Delta S_i$, $i = 1, 2, ..., n$. (This can be done most conveniently by considering a network of intersecting curves of constant $u$ and constant $v$, as shown in Figure 5.30.) Within each segment a suitable point, $(\xi_i, \eta_i, \zeta_i)$, is identified.

With this information, an approximation to the total amount carried by $S$ of the property represented by $f$ can be established:

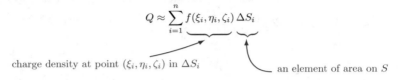

$$Q \approx \sum_{i=1}^{n} f(\xi_i, \eta_i, \zeta_i)\, \Delta S_i$$

charge density at point $(\xi_i, \eta_i, \zeta_i)$ in $\Delta S_i$                    an element of area on $S$

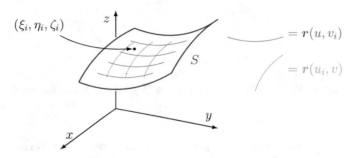

$(\xi_i, \eta_i, \zeta_i)$

$= r(u, v_i)$

$= r(u_i, v)$

**Figure 5.30**  A partition of the surface $S$ and an approximation to $Q$.

We can either assume that $f$ is constant over $\Delta S_i$ or adopt a mean-value-theorem argument.

As we have done many times before (see Section 4.A), we refine the partition into smaller and smaller segments and take the limit as $\Delta S_i \to 0$ and $n \to \infty$. Then, provided the limit exists the result is what is designated to be the surface integral of $f$ over $S$:

$$\iint_S f(x, y, z)\, \mathrm{d}S$$

— the limit exists if $f$ is continuous and $S$ is bounded (Theorem 3.2).

**Remarks**

* It is easily established that (Corollary 4.1.6) if $S = S_1 \cup \cdots \cup S_n$ then

$$\iint_S f\, \mathrm{d}S = \sum_{i=1}^{N} \iint_{S_i} f\, \mathrm{d}S.$$

This is a useful result if $S$ needs partitioning into parts to ensure the existence of one-to-one mappings of those parts onto suitable regions of $\mathbb{R}^2$, and the integral over $S$ considered piecewise.

* If $f = 1$ then, as with Corollary 4.1.3, $\displaystyle\iint_S f \mathrm{d}S = \iint \mathrm{d}S = $ area of $S$.

## Evaluating a surface integral

### • General surface parametrization

To evaluate the surface integral, assuming it is tractable, requires rewriting the area element $\mathrm{d}S$ in terms of known quantities that define the surface. For example, recall from Section 5.A, Page 234, that for a surface parameterized by $u$ and $v$, we can define tangent vectors $\dfrac{\partial \boldsymbol{r}}{\partial u}$ and $\dfrac{\partial \boldsymbol{r}}{\partial v}$ to constant $v$ and constant $u$ curves, respectively. Then, provided $\dfrac{\partial \boldsymbol{r}}{\partial u} \neq 0$, $\dfrac{\partial \boldsymbol{r}}{\partial v} \neq 0$, meaning that $\boldsymbol{r}(u, v_0)$ and $\boldsymbol{r}(u_0, v)$ are smooth curves, and $S$ is a *smooth* surface, the

tangent vectors $\begin{cases} \dfrac{\partial \boldsymbol{r}}{\partial u} \\[2mm] \dfrac{\partial \boldsymbol{r}}{\partial v} \end{cases}$ lead to differential line elements: $\begin{cases} \dfrac{\partial \boldsymbol{r}}{\partial u}\,\mathrm{d}u \\[2mm] \dfrac{\partial \boldsymbol{r}}{\partial v}\,\mathrm{d}v \end{cases}$ and

their cross product leads to the differential area element:
$\mathrm{d}S = \left| \dfrac{\partial \boldsymbol{r}}{\partial u} \times \dfrac{\partial \boldsymbol{r}}{\partial v} \right| \mathrm{d}u\,\mathrm{d}v.$

Consequently, we arrive at a pragmatic representation of the surface integral as a double integral over the planar domain, $D$, that defines the original $S$ ($\mathrm{d}A = \mathrm{d}u\,\mathrm{d}v$).

$$Q = \underbrace{\iint_S f(\boldsymbol{r})\mathrm{d}S}_{\substack{\text{surface integral} \\ \text{of } f \text{ over } S \\ \text{— DEFINITION}}} = \underbrace{\iint_D f\big(\boldsymbol{r}(u,v)\big)\left|\frac{\partial \boldsymbol{r}}{\partial u} \times \frac{\partial \boldsymbol{r}}{\partial v}\right| \mathrm{d}A}_{\substack{\text{double integral of } f.\left|\cdots \times \cdots\right| \text{ over } D \\ \text{leading to an iterated integral} \\ \text{— PRACTICE}}}$$

✍ **Mastery Check 5.13:**

For a general $C^1$ parametrization of a surface $S : \boldsymbol{r} = \boldsymbol{r}(u,v)$, derive the expression for $\dfrac{\partial \boldsymbol{r}}{\partial u} \times \dfrac{\partial \boldsymbol{r}}{\partial v}$.

What then is the expression for $\mathrm{d}S$?

✍

✍ **Mastery Check 5.14:**

Determine the areal moment of inertia about the $z$-axis of the parametric surface $S$ given by

$$x = 2uv, \quad y = u^2 - v^2, \quad z = u^2 + v^2; \quad u^2 + v^2 \le 1.$$

That is, evaluate the integral $\displaystyle\iint_S (x^2 + y^2)\mathrm{d}S$.

✍

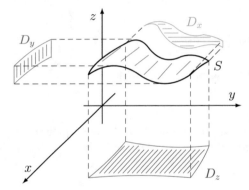

(a) $\boldsymbol{r} = \big(x, y, q(x,y)\big)$
     for the integral over $D_z$

(b) $\boldsymbol{r} = \big(x, g(x,z), z\big)$
     for the integral over $D_y$

(c) $\boldsymbol{r} = \big(h(y,z), y, z\big)$
     for the integral over $D_x$

**Figure 5.31** Projections of $S$ onto the three coordinate planes.

• **Cartesian coordinate representation**

The one and same surface *can* be parameterized, at least piecewise, in any of three different ways with respect to different pairwise combinations of the Cartesian coordinates, as shown in Figure 5.31.

From these different representations we have infinitesimal surface elements shown in Figure 5.32 below.

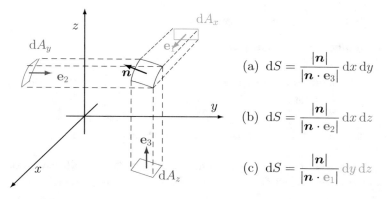

(a) $\mathrm{d}S = \dfrac{|\boldsymbol{n}|}{|\boldsymbol{n} \cdot \mathbf{e}_3|} \, \mathrm{d}x \, \mathrm{d}y$

(b) $\mathrm{d}S = \dfrac{|\boldsymbol{n}|}{|\boldsymbol{n} \cdot \mathbf{e}_2|} \, \mathrm{d}x \, \mathrm{d}z$

(c) $\mathrm{d}S = \dfrac{|\boldsymbol{n}|}{|\boldsymbol{n} \cdot \mathbf{e}_1|} \, \mathrm{d}y \, \mathrm{d}z$

**Figure 5.32**   Projections of a surface area element $\mathrm{d}S$ onto the three coordinate planes.

### ✎ Mastery Check 5.15:

Evaluate the surface area element $\mathrm{d}S$ in the three cases, involving, respectively, the functions $q$, $g$, and $h$, as described in Figure 5.31.

✎

## II. Surface integrals of vector fields

The preceding discussion on surface integrals of scalar functions can be extended directly to surface integrals of vector fields to give new vector quantities. Suppose $\boldsymbol{f} : \mathbb{R}^3 \longrightarrow \mathbb{R}^3$ is a vector field restricted to a smooth surface $S$. Then the surface integral of $\boldsymbol{f}$ over $S$ is simply

$$\iint_S \boldsymbol{f}(x,y,z)\mathrm{d}S = \mathbf{i} \iint_S f_1(x,y,z)\mathrm{d}S + \mathbf{j} \iint_S f_2(x,y,z)\mathrm{d}S + \mathbf{k} \iint_S f_3(x,y,z)\mathrm{d}S.$$

That is, by appealing to the linearity properties of vectors and of the integral operator one can determine the surface integral of a vector field as a vector of surface integrals of the components of that field. However, the following, more important variant of surface integrals of vector fields is the more usual.

One of the most useful qualities of vector fields is that of being able to describe collective movement, in terms of both direction and magnitude of motion. When things are moving, it is often of interest to know how much passes through a given region or across a given area. For example,

- vehicular traffic through an area of a city or an entire city,

- water through a semi-permeable membrane, such as a plant cell wall,

- magnetic flux through a steel cooking pot, or

- X-ray photons through a body.

This "how much" is called the *flux*. To evaluate this quantity we need two pieces of information:

- a vector description of the collective movement — that would be $\boldsymbol{f}$.

- a vector description of the region through which the collective movement is to pass — the surface $(S)$ and the normal to the surface $(\boldsymbol{N})$.

Let's start with a simple situation. Suppose $\boldsymbol{f}$ is a constant vector field with $|\boldsymbol{f}|$ giving the number of photons per unit area, travelling in a constant direction, $\dfrac{\boldsymbol{f}}{|\boldsymbol{f}|}$. Now suppose we wanted to know the number of photons entering a body. A somewhat simple picture of the situation is shown in Figure 5.33.

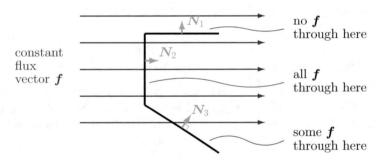

**Figure 5.33** Constant flux $\boldsymbol{f}$ and the boundary of a body.

We see from Figure 5.33 that to determine how much enters a body we need to know both the surface of the body and its relation to the uniform $\boldsymbol{f}$. For a flat surface of area $A$ and *unit normal* $\boldsymbol{N}$ the number of X-ray photons passing through will depend on the area $A$ as well as the latter's orientation with respect to the direction of $\boldsymbol{f}$.

If the vector $\boldsymbol{f}$ makes an angle $\theta$ with the surface normal $\boldsymbol{N}$, then we have

$$\text{flux} \ = |\boldsymbol{f}|\cos\theta\, A \ = \ (\boldsymbol{f}\cdot\boldsymbol{N})A \ = \ \boldsymbol{f}\cdot\boldsymbol{A}.$$

Notice and remember that $\boldsymbol{N}$ must be a unit normal! Why? Because the only feature that is needed is the cosine of the angle between $\boldsymbol{f}$ and $\boldsymbol{A}$; the scalar product of two vectors involves the magnitudes of *both* vectors, but the flux only requires the magnitude of $\boldsymbol{f}$, which will be the case using the scalar product provided $|\boldsymbol{N}| = 1$. Unless otherwise stated in the remainder, $\boldsymbol{N}$ will denote a unit normal vector.

**Remarks**

  * By combining $A$ with $\boldsymbol{N}$ we have made the surface area a vector, $\boldsymbol{A}$.

  * If $\boldsymbol{f} \cdot \boldsymbol{A} = -|\boldsymbol{f}||\boldsymbol{A}|$, we say the $\boldsymbol{f}$ is *into* $\boldsymbol{A}$.

  * If $\boldsymbol{f} \cdot \boldsymbol{A} = |\boldsymbol{f}||\boldsymbol{A}|$, we say the $\boldsymbol{f}$ is *out of* $\boldsymbol{A}$.

The above discussion assumes a constant field and planar surfaces — $\boldsymbol{f}$ does not depend on position, and the surface normals are constant vectors. For non-uniform fields and smooth varying surfaces, Figure 5.34, the same reasoning can be applied, but locally. Suppose again that $\boldsymbol{f} : \mathbb{R}^3 \longrightarrow \mathbb{R}^3$ is defined on a surface $S \subset \mathbb{R}^3$, and the surface is parameterized with respect to parameters $(u, v)$:
$$S = \{\boldsymbol{r}(u, v) : (u, v) \in D \subset \mathbb{R}^2\}.$$

Let $\mathrm{d}S$ be a differential element of area defined at the point $\boldsymbol{r} \in S$, with unit normal $\boldsymbol{N}$; $\mathrm{d}S$ is sufficiently small (infinitesimal, even) that defined over it $\boldsymbol{f}$ is uniform. Locally at $\boldsymbol{r} \in S$ the flux of $\boldsymbol{f}$ through $\mathrm{d}S$ is
$$\boldsymbol{f}(\boldsymbol{r}) \cdot \boldsymbol{N}(\boldsymbol{r}) \, \mathrm{d}S = \boldsymbol{f}(\boldsymbol{r}) \cdot \mathrm{d}\boldsymbol{S}.$$

Accumulating such contributions over the entire surface we get
$$\text{total flux} = \iint_S (\boldsymbol{f} \cdot \boldsymbol{N}) \, \mathrm{d}S = \iint_S \boldsymbol{f} \cdot \mathrm{d}\boldsymbol{S}$$

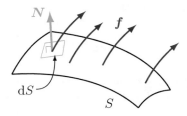

**Figure 5.34** Non-constant $\boldsymbol{f}$ and curved $S$.

The integral $\iint_S (\boldsymbol{f} \cdot \boldsymbol{N})\,\mathrm{d}S$ is in the form of a surface integral of a scalar field, $\boldsymbol{f} \cdot \boldsymbol{N}$. This means that we can use the different approaches to surface integrals discussed on Pages 263 – 264 to evaluate this integral. Two of these means are given in the Remarks below.

**Remarks**

* If the surface $S$ is defined by a level set equation (Section 1.F), that is, if

$$S = \{(x, y, z) : \phi(x, y, z) = C\},$$

then

$$\boldsymbol{N} = \frac{\boldsymbol{\nabla}\phi}{|\boldsymbol{\nabla}\phi|}.$$

* If the surface $S$ is defined parametrically as $S = \{\boldsymbol{r}(u, v) : (u, v) \in D\}$, then

$$\boldsymbol{N} = \frac{\dfrac{\partial \boldsymbol{r}}{\partial u} \times \dfrac{\partial \boldsymbol{r}}{\partial v}}{\left|\dfrac{\partial \boldsymbol{r}}{\partial u} \times \dfrac{\partial \boldsymbol{r}}{\partial v}\right|}.$$

— $u$ and $v$ *can* be Cartesian coordinates

In this case, recalling the formula for $\mathrm{d}S$ on Page 263 and inserting both expressions in the flux definition:

$$\iint_S \boldsymbol{f} \cdot \boldsymbol{N}\mathrm{d}S = \iint_D \boldsymbol{f} \cdot \underbrace{\frac{\left(\dfrac{\partial \boldsymbol{r}}{\partial u} \times \dfrac{\partial \boldsymbol{r}}{\partial v}\right)}{\left|\dfrac{\partial \boldsymbol{r}}{\partial u} \times \dfrac{\partial \boldsymbol{r}}{\partial v}\right|}}_{\boldsymbol{N}} \underbrace{\left|\dfrac{\partial \boldsymbol{r}}{\partial u} \times \dfrac{\partial \boldsymbol{r}}{\partial v}\right|\,\mathrm{d}u\,\mathrm{d}v}_{\mathrm{d}S},$$

we arrive at the conveniently simpler double integral form,

$$\iint_S \boldsymbol{f} \cdot \boldsymbol{N}\mathrm{d}S = \iint_D \boldsymbol{f} \cdot \left(\frac{\partial \boldsymbol{r}}{\partial u} \times \frac{\partial \boldsymbol{r}}{\partial v}\right)\,\mathrm{d}u\,\mathrm{d}v.$$

* Clearly, in considering a surface integral of a vector field $\boldsymbol{f}$, we presume $\boldsymbol{f}$ to be defined over all of $S$.

* We also assume $S$ is a smooth surface, at least in pieces, so that the existence of a tangent plane at every point of $S$, at least piecewise, implies in turn that at every point there exists a unit normal vector.

* An important feature of this entire discussion concerns surface *orientation*: we assume the surface $S$ is piecewise oriented. For a surface to be *orientable* it must possess a *continuously varying normal*, at least piecewise (Figure 5.35).

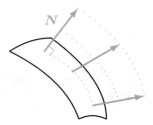

**Figure 5.35**   Continuously turning normal $\boldsymbol{N}(\boldsymbol{r})$.

* From a very practical perspective the choice of parametrization of the surface determines its orientation; a different choice of parametrization can result in the opposite orientation. For example, $\boldsymbol{r}'_u \times \boldsymbol{r}'_v$ determines a specific direction, while $\boldsymbol{r}'_v \times \boldsymbol{r}'_u$ gives the opposite direction.

* By convention the outside of a closed surface is denoted as the positive side, and so we choose a parametrization for a closed surface so that the resulting unit normal vector points to the positive side, i.e. *out* from the region contained within the surface.

For future reference, it is significant to take note that the unit normal $\boldsymbol{N}$ to a (open) surface also specifies the *orientation of the boundary (and vice versa; see Definition 5.11)!* Conventionally, if $\boldsymbol{N}$ satisfies the following condition then the (open) surface $S$ is said to be *positively oriented*.

If the little man in Figure 5.36 walks around the boundary of $S$ so that a vector drawn from his feet to his head points in the same direction as $\boldsymbol{N}$ and $S$ is on the man's left, then $S$ is said to be *positively oriented*.

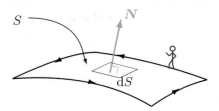

**Figure 5.36**   Mnemonic for determining surface orientation.

## Remarks

* If $S$ is closed we write

$$\oiint_S \boldsymbol{f} \cdot \mathrm{d}\boldsymbol{S} = \oiint_S (\boldsymbol{f} \cdot \boldsymbol{N}) \mathrm{d}S$$

and, as mentioned already, adopt the convention that $\boldsymbol{N}$ is the *out-ward*-pointing unit normal.

* Why is orientation so important? Consider the uniform field $\boldsymbol{f}$ and the open surfaces in Figure 5.37. Of course, for practical purposes we need to choose an $\boldsymbol{N}$ for each surface, and once a choice is made we stick with it. However, the choice of $\boldsymbol{N}$ determines how one interprets the travel of $\boldsymbol{f}$:

If $\boldsymbol{f} \cdot \boldsymbol{N} > 0$ we say $\boldsymbol{f}$ travels *out of* $S$,
if $\boldsymbol{f} \cdot \boldsymbol{N} < 0$ we say $\boldsymbol{f}$ travels *into* $S$.

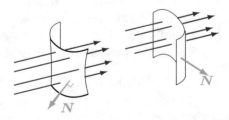

**Figure 5.37**  The relation between surface normal $\boldsymbol{N}$
and a vector field.

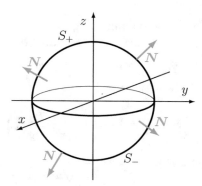

**Figure 5.38**   The sphere of radius $a$ in Example 5.9.

■   **Example 5.9:**

Determine the flux of the vector field $\boldsymbol{f} = (x, y, 2z)$ through the surface $S$, defined by $x^2 + y^2 + z^2 = a^2$.

Note that this surface shown in Figure 5.38 is a closed surface. Hence, we assume the usual convention of taking the *outward*-pointing normal.

We divide the surface into the upper hemisphere,

$$S_+ = \{(x, y, z) : z = g(x, y), 0 \le x^2 + y^2 \le a^2\}$$

with an upward pointing normal, and the lower hemisphere,

$$S_- = \{(x, y, z) : z = -g(x, y), 0 \le x^2 + y^2 \le a^2\}$$

with a downward pointing normal.

Here, $g(x, y) = \sqrt{a^2 - x^2 - y^2}$.

The flux integral over the closed sphere can then be written as the sum of two flux integrals:

$$\oiint_S \boldsymbol{f} \cdot \mathrm{d}\boldsymbol{S} = \iint_{S_+} \boldsymbol{f} \cdot \mathrm{d}\boldsymbol{S} + \iint_{S_-} \boldsymbol{f} \cdot \mathrm{d}\boldsymbol{S}.$$

With the Cartesian variables $x$ and $y$ as parameters, both $S_+$ and $S_-$ are defined through a one-to-one relationship with points in the planar domain $D = \{(x, y) : 0 \le x^2 + y^2 \le a^2\}$.

Hence, the flux integral through $S$ can be rewritten (see Equation (5.2)) as

$$\oiint_S \boldsymbol{f} \cdot \mathrm{d}\boldsymbol{S} = \iint_D \boldsymbol{f}(x, y, g(x, y)) \cdot \left( \frac{\partial \boldsymbol{r}_+}{\partial x} \times \frac{\partial \boldsymbol{r}_+}{\partial y} \right) \mathrm{d}x\, \mathrm{d}y$$

$$+ \iint_D \boldsymbol{f}(x, y, -g(x, y)) \cdot \left( -\frac{\partial \boldsymbol{r}_-}{\partial x} \times \frac{\partial \boldsymbol{r}_-}{\partial y} \right) \mathrm{d}x\, \mathrm{d}y$$

where $r_\pm = (x, y, \pm g(x, y))$, and a minus sign is introduced in the second flux integral to provide the correct surface normal direction.

Now, $\dfrac{\partial r_\pm}{\partial x} = \left(1, 0, \pm \dfrac{\partial g}{\partial x}\right)$ and $\dfrac{\partial r_\pm}{\partial y} = \left(0, 1, \pm \dfrac{\partial g}{\partial y}\right)$.

Thus,

$$\left(\frac{\partial r_\pm}{\partial x} \times \frac{\partial r_\pm}{\partial y}\right) = \left(\mp \frac{\partial g}{\partial x}, \mp \frac{\partial g}{\partial y}, 1\right),$$

where $\dfrac{\partial g}{\partial x} = \dfrac{1}{2}(a^2 - x^2 - y^2)^{-1/2} \cdot (-2x) = \dfrac{-x}{\sqrt{a^2 - x^2 - y^2}} = \dfrac{-x}{g(x, y)}$, etc.

This gives the two normal vectors

$$\pm \left(\frac{\partial r_\pm}{\partial x} \times \frac{\partial r_\pm}{\partial y}\right) = \left(\frac{x}{g}, \frac{y}{g}, \pm 1\right),$$

and so

$$\oiint_S f \cdot dS = \iint_D (x, y, 2g) \cdot \left(\frac{x}{g}, \frac{y}{g}, 1\right) \, dx\, dy$$

$$+ \iint_D (x, y, -2g) \cdot \left(\frac{x}{g}, \frac{y}{g}, -1\right) \, dx\, dy$$

$$= 2 \iint_D \left(\frac{x^2}{g} + \frac{y^2}{g} + 2g\right) dx\, dy = 2 \iint_D \left(\frac{x^2 + y^2 + 2g^2}{g}\right) dx\, dy$$

$$= 2 \iint_D \left(\frac{2a^2 - x^2 - y^2}{\sqrt{a^2 - x^2 - y^2}}\right) dx\, dy.$$

Since $D$ is a disc we can use polar coordinates:

$$x = r \cos\theta, \quad y = r \sin\theta, \quad dx\, dy = r\, dr\, d\theta, \quad x^2 + y^2 = r^2.$$

$$\oiint_S f \cdot dS = 2 \int_0^{2\pi} d\theta \int_0^a \left(\frac{2a^2 - r^2}{\sqrt{a^2 - r^2}}\right) r\, dr$$

$$= 4\pi \left\{ a^2 \int_0^a \frac{r\, dr}{\sqrt{a^2 - r^2}} + \int_0^a \sqrt{a^2 - r^2}\, r\, dr \right\}.$$

The final integrals can be evaluated easily using the substitution $u = a^2 - r^2 \implies du = -2r\, dr$:

$$\oiint_S f \cdot dS = 4\pi \left\{ a^2 \left[ -\frac{1}{2} u^{1/2} \cdot 2 \right]_{a^2}^0 + \left[ -\frac{1}{2} u^{3/2} \cdot \frac{2}{3} \right]_{a^2}^0 \right\}$$

$$= 4\pi \left\{ a^2 \cdot a + a^3 \frac{1}{3} \right\} = \frac{16}{3}\pi a^3.$$

■

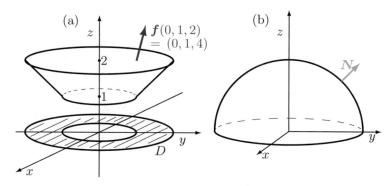

**Figure 5.39**  (a) The cone $S$ and projection $D$ in MC 5.16;
(b) The hemisphere in MC 5.17.

✍  **Mastery Check 5.16:**

Calculate the flux of $\boldsymbol{f} = (x^2, y^2, z^2)$ through the surface

$$S = \{\boldsymbol{r} : z^2 = x^2 + y^2, 1 \leq z \leq 2\}$$

in the direction $\boldsymbol{N} \cdot \mathbf{e}_3 > 0$ (Figure 5.39(a)).

✍

✍  **Mastery Check 5.17:**

Determine the flux of

$\boldsymbol{f} = (-y, x, x^2 + z)$  through  the  surface  $S = \{(x, y, x) : x^2 + y^2 + z^2 = 1,$
$z \geq 0\}$ in Figure 5.39(b).

✍

# 5.E   Gauss's theorem

In this section we consider the topic of the net fluxes through closed surfaces, Figure 5.40.

Consider a $C^1$ field $\boldsymbol{f} : D_f \subset \mathbb{R}^3 \longrightarrow \mathbb{R}^3$ and a *closed* surface $S$ which is the boundary of a *bounded* volume $V \subset D_f$.

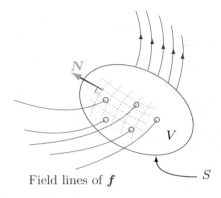

Field lines of $f$

**Figure 5.40** The field $f$ passing into and out of a closed region $S$.

With $S$ being closed there will be some flux of $f$ into $S$ and some flux out of $S$. Consequently, if we take $N$ to be the outward normal to $S$, then $\oiint f \cdot dS$ will be a measure of the *net flux of $f$ out of* $S$:

$$q = \oiint_S f \cdot N \, dS$$

Let's now place this $q$ in a physical setting. Suppose $f$ describes flow of water. Then

- if $q > 0$: this says that more water flows out of $S$ than in, meaning that there is a *production* of water *inside $V$*;

- if $q < 0$: this says that less water flows out of $S$ than in, meaning that there is a *destruction* of water *inside $V$*;

- if $q = 0$: this says that what flows in flows out of $S$, meaning that the amount of water in $V$ is conserved.

From this interpretation one would naturally suspect that $q$ contains information about what occurs *inside $V$*. That is, there is reason to suspect a relationship of the form

$$\iiint_V (\text{production of } f) \, dV = \oiint_s f \cdot dS.$$

This is in fact exactly what Gauss's[1] theorem states:

---

[1]This book uses this form of the possessive for proper names ending in "s", such as Gauss and Stokes [21, 22]. Some texts use the form "Gauss' theorem". The theorems are the same however they are described.

> **Theorem 5.3**
> *Suppose $\boldsymbol{f} : \mathbb{R}^3 \longrightarrow \mathbb{R}^3$ is a $C^1$ vector field defined on and within a domain $V$ which is bounded by a piecewise smooth closed surface $S$ which in turn has a continuously varying outward unit normal $N$. Then*
>
> $$\oiint_S \boldsymbol{f} \cdot \mathrm{d}\boldsymbol{S} = \iiint_V \left( \frac{\partial f_1}{\partial x} + \frac{\partial f_2}{\partial y} + \frac{\partial f_3}{\partial z} \right) \mathrm{d}V = \iiint_V (\boldsymbol{\nabla} \cdot \boldsymbol{f}) \, \mathrm{d}V.$$

**Remarks**

* The theorem implies that the divergence $\boldsymbol{\nabla} \cdot \boldsymbol{f}$ is a measure of the *local production* or *local destruction* of $\boldsymbol{f}$ inside $V$: $\boldsymbol{\nabla} \cdot \boldsymbol{f}(\boldsymbol{x})$ is the *source or sink strength* per unit volume of $\boldsymbol{f}$ at $\boldsymbol{x} \in V$.

* The theorem holds true *only* if $\boldsymbol{N}$ is the unit normal pointing *away* from region $V$, as per the examples in Figure 5.41.

**Figure 5.41**  Two closed surfaces and their outward normals.

* The reference to "piecewise smooth" means that $S$ can have edges, just as long as $S$ is closed.

* Gauss's theorem is useful in rewriting relations involving surface and volume integrals so that all terms can be combined under one integral sign ([1] Chapter 18).

* If we apply the mean value theorem for multiple integrals to the volume integral in Gauss's theorem we get

$$\iiint_V (\boldsymbol{\nabla} \cdot \boldsymbol{f}) \, \mathrm{d}V = \boldsymbol{\nabla} \cdot \boldsymbol{f}(P_0).V,$$

where $\boldsymbol{f} \in C^1$ and is bounded, and where $P_0$ is some point in $V$. If we now take the limit of this result as $V \to 0$ and $S \to 0$ so as to converge to the single point $\boldsymbol{x}$, which will coincide with $P_0$, then

$$\boldsymbol{\nabla} \cdot \boldsymbol{f}(\boldsymbol{x}) = \lim_{V \to 0} \frac{1}{V} \oiint_S \boldsymbol{f} \cdot \mathrm{d}\boldsymbol{S},$$

where $\boldsymbol{x}$ is common to all $V$ and $S$ in this limit.

This result says that the divergence of a vector field $\boldsymbol{f}$ is the *flux per unit volume* of $\boldsymbol{f}$ out of a region of vanishing volume.

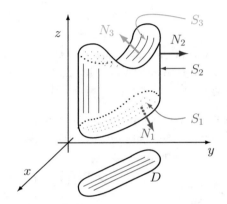

**Figure 5.42** $V$ as a $z$-simple region.

## Sketch proof of Gauss's theorem

We start by splitting the surface and volume integrals into their component terms:

$$\oiint_S \boldsymbol{f} \cdot \boldsymbol{N} \,\mathrm{d}S = \underbrace{\oiint_S (f_1 \mathbf{e}_1) \cdot \mathrm{d}\boldsymbol{S}}_{} + \underbrace{\oiint_S (f_2 \mathbf{e}_2) \cdot \mathrm{d}\boldsymbol{S}}_{} + \underbrace{\oiint_S (f_3 \mathbf{e}_3) \cdot \mathrm{d}\boldsymbol{S}}_{}$$

$$\iiint_V \boldsymbol{\nabla} \cdot \boldsymbol{f} \,\mathrm{d}V = \underbrace{\iiint_V \frac{\partial f_1}{\partial x} \,\mathrm{d}V}_{\text{involves } f_1} + \underbrace{\iiint_V \frac{\partial f_2}{\partial y} \,\mathrm{d}V}_{\text{involves } f_2} + \underbrace{\iiint_V \frac{\partial f_3}{\partial z} \,\mathrm{d}V}_{\text{involves } f_3}.$$

It is always possible to treat $S$ as the union of piecewise smooth surfaces and $V$ as the union of simple domains ($x$-simple, $y$-simple, and $z$-simple).

Suppose now that $V$ is one of these cases, specifically a $z$-simple domain, as in Figure 5.42: $V = \{\boldsymbol{x} : h(x, y) \leq z \leq g(x, y), \ (x, y) \in D\}$.

On $S_1$:   $S_1 = \{\boldsymbol{x} : (x,y) \in D,\ z = h(x,y)\}$,   $\boldsymbol{N}_1 = \dfrac{\left(\dfrac{\partial h}{\partial x}, \dfrac{\partial h}{\partial y}, -1\right)}{\sqrt{\cdots\cdots\cdots}}$.

On $S_2$:   $\boldsymbol{N}_2$ is orthogonal to $\mathbf{e}_3$.

On $S_3$:   $S_3 = \{\boldsymbol{x} : (x,y) \in D,\ z = g(x,y)\}$,   $\boldsymbol{N}_3 = \dfrac{\left(\dfrac{\partial g}{\partial x}, \dfrac{\partial g}{\partial y}, 1\right)}{\sqrt{\cdots\cdots\cdots}}$.

Now let's work with the $f_3$ component of the flux integral.

$$
\oiint_S (f_3\mathbf{e}_3) \cdot \mathrm{d}\boldsymbol{S} = \iint_{S_1 \cup S_2 \cup S_3} (f_3\mathbf{e}_3 \cdot \boldsymbol{N})\,\mathrm{d}S
$$
$$
= \iint_{S_1} (f_3\mathbf{e}_3 \cdot \boldsymbol{N}_1)\,\mathrm{d}S + \iint_{S_2} (f_3\mathbf{e}_3 \cdot \overset{0}{\boldsymbol{N}_2})\,\mathrm{d}S + \iint_{S_3} (f_3\mathbf{e}_3 \cdot \boldsymbol{N}_3)\,\mathrm{d}S
$$
$$
= -\iint_D f_3(x,y,h)\,\mathrm{d}A + \iint_D f_3(x,y,g)\,\mathrm{d}A
$$
$$
= \iint_D \left( \int_{h(x,y)}^{g(x,y)} \frac{\mathrm{d}f_3}{\mathrm{d}z}\,\mathrm{d}z \right) \mathrm{d}A = \iiint_V \frac{\partial f_3}{\partial z}\,\mathrm{d}V.
$$

We can manipulate the other cases in analogous ways. Adding these contributions we get the desired result.   ■

## ■   Example 5.10:

Determine the flux of $\boldsymbol{f} = (x^3, y^3, z^3)$ out of the sphere in Figure 5.38:
$S = \{(x,y,z) : x^2 + y^2 + z^2 = a^2\}$.

The vector field $\boldsymbol{f}$ is clearly $C^1$, and $S$ is closed with a continuously varying normal, $\boldsymbol{N}$. So we can use Gauss's theorem.

$$
\oiint_S \boldsymbol{f} \cdot \mathrm{d}\boldsymbol{S} = \iiint_V \boldsymbol{\nabla} \cdot \boldsymbol{f}\,\mathrm{d}V
$$
$$
= \iiint_V (3x^2 + 3y^2 + 3z^2)\,\mathrm{d}V = 3\iiint_V (x^2 + y^2 + z^2)\,\mathrm{d}V.
$$

We naturally change to spherical coordinates. (See Mastery Check 4.7.)

$$
x = \rho\sin\phi\cos\theta,\ y = \rho\sin\phi\sin\theta,\ z = \rho\cos\phi,\ x^2 + y^2 + z^2 = \rho^2.
$$

The values $0 \le \rho \le a,\ 0 \le \theta \le 2\pi,\ 0 \le \phi \le \pi$ cover all points in $V$, the interior of $S$. The differential $dV = \rho^2 \sin\phi \, d\rho \, d\theta \, d\phi$. Thus

$$\oiint_S \boldsymbol{f} \cdot d\boldsymbol{S} = 3 \iiint_V \left( x^2 + y^2 + z^2 \right) dV$$

$$= 3 \int_0^{2\pi} d\theta \int_0^{\pi} d\phi \int_0^a \rho^2 . \rho^2 \sin\phi \, d\rho = 3 \int_0^{2\pi} d\theta \int_0^{\pi} \frac{1}{5} a^5 \sin\phi \, d\phi$$

$$= \frac{3}{5} a^5 \int_0^{2\pi} d\theta \Big[ - \cos\phi \Big]_0^{\pi} = \frac{12\pi a^5}{5}.$$

∎

**What is good about Gauss's theorem:**

Assuming that the conditions of the theorem are satisfied, if we were asked to evaluate $\iiint_V g(x, y, z) \, dV$, we could instead evaluate $\oiint_S \boldsymbol{f} \cdot d\boldsymbol{S}$, where $\boldsymbol{\nabla} \cdot \boldsymbol{f} = g(x, y, z)$. On the other hand, if we were asked to evaluate $\oiint_S \boldsymbol{f} \cdot d\boldsymbol{S}$, we could instead evaluate $\iiint_V g(x, y, z) \, dV$.

In these cases we choose whichever integral is easiest to evaluate.

■   **Example 5.11:**

According to Corollary 4.1.3, volumes of regions are found when $g = 1$. To get $g = 1$, use $\boldsymbol{f} = \frac{1}{3}\boldsymbol{r} = \frac{1}{3}(x, y, z)$, or $= (x, 0, 0)$, or $= (0, y, 0)$.

∎

**What is NOT so good about Gauss's theorem:**

The conditions on the theorem are strict. As given, the theorem states that

$$\oiint_S \boldsymbol{f} \cdot d\boldsymbol{S} = \iiint_V \boldsymbol{\nabla} \cdot \boldsymbol{f} \, dV$$

is true *only if*

a) $S$ is a closed surface. But not every problem that is posed involves a closed surface.

However, if the given $S$ in a problem is not closed we can create a convenient closed surface by complementing $S$:

$$S \longrightarrow S_c = S \cup S_{\mathrm{xtra}}$$

$$\underset{\text{closed}}{\nearrow} \qquad \underset{\text{extra bit}}{\nearrow}$$

and then apply Gauss's theorem on $S_c$ since, by Corollary 5.2.1, we can argue that

$$\iint_S \boldsymbol{f} \cdot \mathrm{d}\boldsymbol{S} = \oiint_{S_c} \boldsymbol{f} \cdot \mathrm{d}\boldsymbol{S} - \iint_{S_{\mathrm{xtra}}} \boldsymbol{f} \cdot \mathrm{d}\boldsymbol{S} = \iiint_V \boldsymbol{\nabla} \cdot \boldsymbol{f}\, \mathrm{d}V - \iint_{S_{\mathrm{xtra}}} \boldsymbol{f} \cdot \mathrm{d}\boldsymbol{S}.$$

— by Gauss's theorem on $S_c$
at the middle step

(See Mastery Check 5.18.)

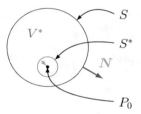

**Figure 5.43**  Exclusion of the singular point $P_0$.

b) $\boldsymbol{f}$ is $C^1$ on $S$ and in $V$. However, in some problems, $\boldsymbol{f}$ may be defined on $S$ so that we can evaluate a flux integral, but not defined at some point $P_0$ within $S$, so we cannot use Gauss's theorem.

However, we can always *exclude* $P_0$ by enclosing it with another surface. For example, suppose $S$ is closed and is the boundary of $V$ containing the singular point $P_0$. Enclose $P_0$ in a new closed surface $S^*$, as shown in Figure 5.43.

Now $S \cup S^*$ is the boundary of a volume $V^*$, and $P_0 \notin V^*$. Hence, with the singular point now removed we can argue, again by appealing to Corollary 5.2.1, that

$$\oiint_S \boldsymbol{f} \cdot \mathrm{d}\boldsymbol{S} = \oiint_{S \cup S^*} \boldsymbol{f} \cdot \mathrm{d}\boldsymbol{S} - \oiint_{S^*} \boldsymbol{f} \cdot \mathrm{d}\boldsymbol{S} = \iiint_{V^*} \boldsymbol{\nabla} \cdot \boldsymbol{f}\, \mathrm{d}V - \oiint_{S^*} \boldsymbol{f} \cdot \mathrm{d}\boldsymbol{S}.$$

— by Gauss's theorem on $S \cup S^*$
at the middle step

(See Mastery Check 5.20.)

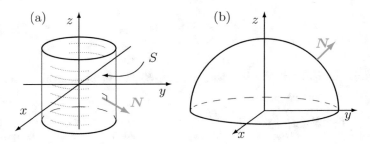

**Figure 5.44**  (a) The cylinder $x^2 + y^2 = 1$ in MC 5.18;
(b) The unit hemisphere in MC 5.19.

### ✍ Mastery Check 5.18:
**How to use Gauss's theorem when $f$ is $C^1$ but $S$ is not closed.**

Determine the flux of $f = (xz^2, x, x^2)$ through the cylinder
$S = \{x : x^2 + y^2 = 1, \quad |z| \leq 1\}$ with outward-pointing normal (Figure 5.44(a)).

✍

### ✍ Mastery Check 5.19:

Use Gauss's theorem to find the flux of the vector field $f = (-y, x, x^2 + z)$
through the surface $S = \{(x, y, z) : x^2 + y^2 + z^2 = 1, \ z \geq 0\}$ (Figure 5.44(b)).

✍

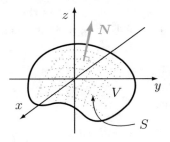

**Figure 5.45**  An arbitrary closed $S$ enclosing $(0, 0, 0)$.

✎   **Mastery Check 5.20:**
**How to use Gauss's theorem if $S$ is closed but $f$ is not $C^1$ at some point.**

Determine the flux of $f = -k\dfrac{r}{|r|^3}$ with $r = (x, y, z)$, through *any* closed surface, Figure 5.45, enclosing a volume $V$ which contains the point $(0, 0, 0)$.

✎

✎   **Mastery Check 5.21:**

Show that for the conditions of Gauss's theorem and for two $C^2$ functions $\phi, \psi : \mathbb{R}^3 \longrightarrow \mathbb{R}$, the following integral identity holds:

$$\iint_V \left(\psi \boldsymbol{\nabla}^2 \phi - \phi \boldsymbol{\nabla}^2 \psi\right) \mathrm{d}V = \oiint_S \left(\psi \boldsymbol{\nabla} \phi - \phi \boldsymbol{\nabla} \psi\right) \cdot \boldsymbol{N} \mathrm{d}S.$$

This is known as Green's second identity ([5]) or Green's formula ([13]).

✎

# 5.F   Green's and Stokes's theorems

Returning to the subject of line integrals of vector fields, but focusing interest on closed curves, we come to discuss Green's theorem and Stokes's theorem, which do for circulation of vector fields (Page 255) what Gauss's theorem does for net fluxes (Page 273). However, before introducing the theorems we shall first cover a few additional curve concepts.

**I. Additional notes on curves**

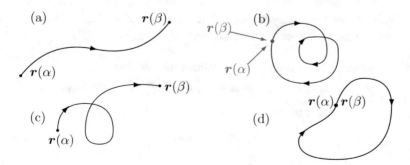

**Figure 5.46**  Curve types.

(i) For any smooth curve (Figure 5.46(a)) we can find or construct a *parametrization* $\boldsymbol{r} = \boldsymbol{r}(t)$, $\alpha \le t \le \beta$.

(ii) If $\boldsymbol{r}(\alpha) = \boldsymbol{r}(\beta)$ (Figure 5.46(b)), then the curve is *closed*.

(iii) If $\boldsymbol{r}(t_1) = \boldsymbol{r}(t_2)$ for some $\alpha < t_1 < t_2 < \beta$ (Figure 5.46(c)), the curve is *self-intersecting*.

(iv) If $\boldsymbol{r}(t_1) = \boldsymbol{r}(t_2)$ for some $\alpha \le t_1 < t_2 \le \beta \implies t_1 = \alpha, t_2 = \beta$ (Figure 5.46(d)), the curve is closed and non-intersecting. Such a curve is called *a simple closed curve*.

(v) Of crucial importance is the obvious fact that a simple closed curve in 3D can be the boundary of many open surfaces, in 3D, as indicated by the three example surfaces in Figure 5.47.

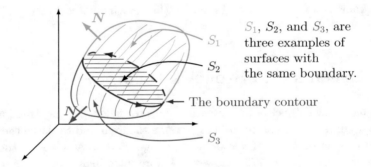

$S_1$, $S_2$, and $S_3$, are three examples of surfaces with the same boundary.

The boundary contour

**Figure 5.47** One contour, multiple surfaces.

So, in 3D, while an *open* surface has a unique boundary, which is a *simple closed curve*, the converse is not true. In 3D, a *closed* curve can be the boundary to an infinite number of surfaces. This becomes important in the context of Stokes's theorem (Page 287).

(vi) A parametrization of a curve defines its orientation:

$$t \; : a \longrightarrow b \quad \implies \quad \boldsymbol{r}_0 = \boldsymbol{r}(a) \longrightarrow \boldsymbol{r}(b) = \boldsymbol{r}_1$$
$$t^* : \alpha \longrightarrow \beta \quad \implies \quad \boldsymbol{r}_1 = \boldsymbol{r}(\alpha) \longrightarrow \boldsymbol{r}(\beta) = \boldsymbol{r}_0.$$

**Figure 5.48** One non-closed curve, two alternative directions.

There are two alternatives to specifying orientation. The choice of alternative is obvious for a non-closed curve: Figure 5.48. For a closed curve we need to be more precise, which the next definition and Figure 5.49 attempt to address.

**Definition 5.11**

*A closed curve* $\Gamma$, *the boundary of an* oriented surface $S$, *has* positive orientation *if*

*(a)* 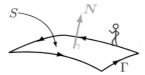 *where* $N$ *is a unit surface normal, and*

*(b) as we walk around* $\Gamma$ *the surface is on our left.*

**Figure 5.49** A positively oriented boundary.

■ **Example 5.12:**

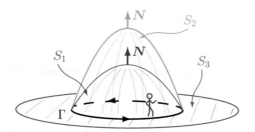

**Figure 5.50** Surface and boundary orientation.

$\Gamma$ in Figure 5.50 is a positively oriented boundary to both $S_1$ and $S_2$, but not $S_3$! ■

## II. Green's theorem

In the case of $\mathbb{R}^2$, the ambiguity between a closed contour and the surface it encloses vanishes; a closed positively oriented curve defines and encloses a unique region (Figure 5.51). We take advantage of that fact in the next theorem which is valid for plane vector fields and plane regions (in 2D life is much simpler).

We state and give a sketch proof of this important result, which was discovered by a gentleman having no formal education at the time [18].

---

**Theorem 5.4**

*Green's theorem:*

*Let $\boldsymbol{f}(x,y) = f_1(x,y)\mathbf{e}_1 + f_2(x,y)\mathbf{e}_2$ be a smooth 2D field defined on and within the positively oriented simple closed boundary $\Gamma$ of a closed and bounded region $D \subset \mathbb{R}^2$. Then*

$$\oint_\Gamma \boldsymbol{f} \cdot \mathrm{d}\boldsymbol{r} = \oint_\Gamma f_1(x,y)\,\mathrm{d}x + f_2(x,y)\,\mathrm{d}y = \iint_D \left(\frac{\partial f_2}{\partial x} - \frac{\partial f_1}{\partial y}\right) \mathrm{d}A.$$

$\underbrace{\qquad}$ *line integral* $\qquad\qquad\qquad$ *double integral* $\underbrace{\qquad}$

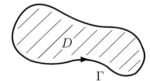

**Figure 5.51** *Domain $D$ and its positively oriented boundary $\Gamma$.*

---

**Sketch proof of Green's theorem (for a simply-connected domain)**

Suppose $D$ is $x$-simple and $y$-simple. As indicated in Figure 5.52 functions $\phi_1$, $\phi_2$, $\psi_1$, and $\psi_2$ can be found such that

$$D = \{(x,y) : a \le x \le b, \ \phi_1(x) \le y \le \phi_2(x)\}$$

and

$$D = \{(x,y) : c \le y \le d, \ \psi_1(y) \le x \le \psi_2(y)\}.$$

Suppose also that all integrals exist.

We shall show that $\displaystyle\iint_D \frac{\partial f_2}{\partial x}\,\mathrm{d}A = \oint_\Gamma f_2\,\mathrm{d}y$:

Treating $D$ as a $y$-simple domain we get

$$\iint_D \frac{\partial f_2}{\partial x}\, dA = \int_c^d dy \int_{\psi_1(y)}^{\psi_2(y)} \frac{\partial f_2}{\partial x}\, dx$$

$$= \int_c^d f_2(\psi_2(y), y)\, dy - \int_c^d f_2(\psi_1(y), y)\, dy$$

$$= \int_{\psi_2} f_2\, dy + \int_{-\psi_1} f_2\, dy \;=\; \oint_\Gamma f_2\, dy.$$

An analogous argument treating $D$ as an $x$-simple domain will show that $\iint_D \frac{\partial f_1}{\partial y}\, dA = -\oint_\Gamma f_1\, dx$, and the theorem is proved.

∎

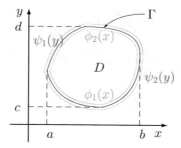

**Figure 5.52**  $D$ as an $x$-simple and $y$-simple domain.

## ■   Example 5.13:

Use Green's theorem to evaluate a line integral.

Calculate $\oint_\Gamma \boldsymbol{f}(\boldsymbol{r}) \cdot d\boldsymbol{r}$ where $\Gamma$ is the boundary of the rectangle $D = \{(x, y) : x \in [0, 4], y \in [0, 1]\}$ when $\boldsymbol{f} = 3x^2\mathbf{i} - 4xy\mathbf{j}$.

We have

$$\oint_\Gamma \boldsymbol{f}(\boldsymbol{r}) \cdot d\boldsymbol{r} = \iint_D \left( \frac{\partial f_2}{\partial x} - \frac{\partial f_1}{\partial y} \right) dA$$

$$= \int_0^4 dx \int_0^1 (-4y - 0)\, dy \;=\; \int_0^4 (-2)\, dx \;=\; -8.$$

■

## ✍  Mastery Check 5.22:

For the function $\boldsymbol{f}$ and curve $\Gamma$ as defined in Example 5.13, find $\oint_\Gamma \boldsymbol{f}(\boldsymbol{r}) \cdot d\boldsymbol{r}$

as a line integral, i.e. without using Green's theorem.

✍

## ✍ Mastery Check 5.23:

Verify Green's theorem for the curve integral

$$\oint_\Gamma (3x^2 - 8y^2)\,\mathrm{d}x + (4y - 6xy)\,\mathrm{d}y$$

where $\Gamma$ is the boundary of the region $D$ bounded by curves $y = \sqrt{x}$ and $y = x^2$ shown in Figure 5.53.                                                    ✍

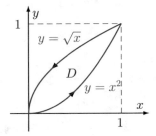

**Figure 5.53**  $D$ and its boundary: $y = \sqrt{x}$ and $y = x^2$.

## Remarks

* As with Gauss's theorem, this theorem asserts a relation between properties of $f$ within a region and other properties on the boundary!

* Special choices of 2D fields allow for the calculation of areas of planar regions using curve integrals, as in the next Mastery Check.

### ✍ Mastery Check 5.24:

Prove the preceding remark for each of the following cases. $F = x\mathbf{e}_2$;   $F = -y\mathbf{e}_1$;   $F = \frac{1}{2}(-y\mathbf{e}_1 + x\mathbf{e}_2)$.
Hint: Recall Corollary 4.1.3 which gives a formula for area, and compare this with the statement of Green's theorem.

✍

* Green's theorem is valid for more complicated regions. For example, it works in the case of annular domains such as shown in Figure 5.54. Take good note of the indicated orientations of the inner and outer boundaries of the annular domain. In each case the curve complies with Definition 5.11 so that Green's theorem remains valid.

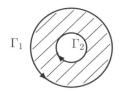

**Figure 5.54**  The composite boundary $\Gamma = \Gamma_1 \cup \Gamma_2$.

* If the plane field $\boldsymbol{f}$ is conservative then

$$\oint_\Gamma \boldsymbol{f} \cdot \mathrm{d}\boldsymbol{r} = 0 \quad \Longrightarrow \quad \iint_D \left( \frac{\partial f_2}{\partial x} - \frac{\partial f_1}{\partial y} \right) \mathrm{d}A = 0,$$

which motivates the following important theorem about conservative fields.

---

**Theorem 5.5**

*If the plane field $\boldsymbol{f} = (f_1, f_2)$ satisfies $\dfrac{\partial f_2}{\partial x} = \dfrac{\partial f_1}{\partial y}$ in a simply-connected domain $D$ then $\boldsymbol{f}$ is conservative in $D$.*

---

### III. Stokes's theorem

Generalizing Green's theorem to 3D we get what is commonly referred to as Stokes's theorem, which analogously relates properties of a vector field around a boundary of an (unclosed) surface to properties of the field (in this case the curl of the field) all over that surface.

---

**Theorem 5.6**

*Suppose $S$ is an oriented surface, piecewise smooth in $\mathbb{R}^3$, with unit normal $\boldsymbol{N}$, and suppose that $S$ is bounded by a piecewise smooth closed curve $\Gamma$ with positive orientation. If $\boldsymbol{f} = \big(f_1(x,y,z), f_2(x,y,z), f_3(x,y,z)\big)$ is a 3D smooth field defined on $S$, then*

$$\oint_\Gamma \boldsymbol{f} \cdot \mathrm{d}\boldsymbol{r} = \iint_S (\boldsymbol{\nabla} \times \boldsymbol{f}) \cdot \boldsymbol{N} \, \mathrm{d}S.$$

---

**Remarks**

* Suppose we apply the mean value theorem to the surface integral in Stokes's theorem:

$$\iint_S (\boldsymbol{\nabla} \times \boldsymbol{f}) \cdot \mathrm{d}\boldsymbol{S} = (\boldsymbol{\nabla} \times \boldsymbol{f})\Big|_{\boldsymbol{r}_0} \cdot \boldsymbol{N}(\boldsymbol{r}_0) \iint_S \mathrm{d}S$$

for $\boldsymbol{f} \in C^1$ and some point $\boldsymbol{r}_0 \in S$. In the limit, as $S, \Gamma \longrightarrow 0$

$$(\boldsymbol{\nabla} \times \boldsymbol{f})\Big|_{\boldsymbol{r}} \cdot \boldsymbol{N}(\boldsymbol{x}) = \lim_{S \to 0} \frac{1}{S} \oint_\Gamma \boldsymbol{f} \cdot \mathrm{d}\boldsymbol{r} \ (\boldsymbol{x} \text{ common to all } S \text{ in this limit})$$

> — the component of $\boldsymbol{\nabla} \times \boldsymbol{f}$ normal to $S$ at $\boldsymbol{x} \in S$
> is the work done per unit area in traversing
> an oriented contour $\Gamma$ of
> vanishing length

* If $\boldsymbol{f} \in C^1(\mathbb{R}^3)$ is conservative, then by Theorem 5.2 $\oint_\Gamma \boldsymbol{f} \cdot \mathrm{d}\boldsymbol{r} = 0$, and Stokes's theorem implies that $\iint_S (\boldsymbol{\nabla} \times \boldsymbol{f}) \cdot \mathrm{d}\boldsymbol{S} = 0$.

* Conversely, if for $\boldsymbol{f} \in C^1(\mathbb{R}^3)$ we find that

$$\frac{\partial f_1}{\partial y} = \frac{\partial f_2}{\partial x}, \quad \frac{\partial f_3}{\partial x} = \frac{\partial f_1}{\partial z}, \quad \frac{\partial f_3}{\partial y} = \frac{\partial f_2}{\partial z}$$

in some simply-connected domain, then

$$\operatorname{curl} \boldsymbol{f} = \boldsymbol{\nabla} \times \boldsymbol{f} = 0.$$

> — the vector field $\boldsymbol{f}$ is then said to be irrotational

and Stokes's theorem implies that $\oint_\Gamma \boldsymbol{f} \cdot \mathrm{d}\boldsymbol{r} = 0$ for all closed curves $\Gamma$ in that domain. That is, $\boldsymbol{f}$ is *conservative*. This important result is an extension to 3D of Theorem 5.5 and warrants its own theorem status.

---

**Theorem 5.7**
*If the $C^1$ vector field $\boldsymbol{f} = (f_1, f_2, f_3)$ satisfies $\boldsymbol{\nabla} \times \boldsymbol{f} = 0$ in a simply-connected domain $D \subset \mathbb{R}^3$, then $\boldsymbol{f}$ is conservative in $D$.*

---

■ **Example 5.14:**
Verify Stokes's theorem for the $C^1$ vector field $\boldsymbol{f} = (xz, xy^2 + 2z, xy + z)$,

where the contour $\Gamma$ is created by the intersection of the cone $x = 1 - \sqrt{y^2 + z^2}$ and the $xy$- and $yz$-planes, such that $0 \leq x \leq 1$, $-1 \leq y \leq 1$, $z \geq 0$. $\Gamma$ is oriented counterclockwise as seen from $(0, 0, 10)$.

**Solution:** The geometrical situation is shown in Figure 5.55.

The piecewise curve $\Gamma$ comprises three part curves: $\Gamma_1$, $\Gamma_2$, $\Gamma_3$, which are oriented as shown.

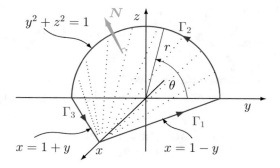

**Figure 5.55** The cone $x = 1 - \sqrt{y^2 + z^2}$.

Note that no surface with which to verify the theorem is specified. However, two possibilities come to mind: The curved surface of the cone lying above the $xy$-plane, and the piecewise combination of the two planar pieces of the $xy$-plane and the $yz$-plane within the bottom triangle and semicircle, respectively.

First, the curve integral, $\displaystyle\oint_\Gamma \boldsymbol{f} \cdot d\boldsymbol{r} = \left( \int_{\Gamma_1} + \int_{\Gamma_2} + \int_{\Gamma_3} \right) \boldsymbol{f} \cdot d\boldsymbol{r}$, where

$$\int_{\Gamma_1} \boldsymbol{f} \cdot d\boldsymbol{r} = \int_{\Gamma_1} xz\,dx + \left(xy^2 + 2z\right)dy + \left(xy + z\right)dz = \int_{\Gamma_1} xy^2\,dy \quad z, dz = 0$$

$$= \int_0^1 (1-y)y^2\,dy = \frac{1}{3} - \frac{1}{4} = \frac{1}{12},$$

and, with $x = 0$, $y = \cos\theta$, $z = \sin\theta$,

$$\int_{\Gamma_2} \boldsymbol{f} \cdot d\boldsymbol{r} = \int_0^\pi \boldsymbol{f}(\boldsymbol{r}(\theta)) \cdot \frac{d\boldsymbol{r}}{d\theta}\,d\theta = \int_0^\pi \left( (2\sin\theta)(-\sin\theta) + \sin\theta\cos\theta \right)d\theta$$

$$= -2 \int_0^\pi \sin^2\theta\,d\theta + \frac{1}{2} \int_0^\pi \sin 2\theta\,d\theta = -2.\frac{1}{2}.\pi = -\pi$$

$$\int_{\Gamma_3} \boldsymbol{f} \cdot d\boldsymbol{r} = \int_{\Gamma_3} xy^2\,dy = \int_{-1}^0 (1+y)y^2\,dy = \frac{1}{3} - \frac{1}{4} = \frac{1}{12}.$$

Therefore, adding these contributions we get $\displaystyle\int_\Gamma \boldsymbol{f}\cdot\mathrm{d}\boldsymbol{r} \;=\; \frac{1}{12}-\pi+\frac{1}{12}\;=\;$
$\dfrac{1}{6}-\pi.$

Now consider the surface integral, $\displaystyle\iint_S (\boldsymbol{\nabla}\times\boldsymbol{f})\cdot\boldsymbol{N}\,\mathrm{d}S.$

The correct orientation of the cone surface is with $\boldsymbol{N}$ pointing to $+$ve $z$.

$$\boldsymbol{\nabla}\times\boldsymbol{f} \;=\; \begin{vmatrix} \mathbf{i} & \mathbf{j} & \mathbf{k} \\[4pt] \dfrac{\partial}{\partial x} & \dfrac{\partial}{\partial y} & \dfrac{\partial}{\partial z} \\[6pt] xz & xy^2+2z & xy+z \end{vmatrix} \;=\; (x-2)\mathbf{i}+(x-y)\mathbf{j}+y^2\mathbf{k}.$$

Consider the parametrization for a cone
$x = 1 - r,\ y = r\cos\theta,\ z = r\sin\theta;\ D = \{0 \le \theta \le \pi,\ 0 \le r \le 1\}.$

$$\frac{\partial \boldsymbol{x}}{\partial r}\times\frac{\partial \boldsymbol{x}}{\partial\theta} \;=\; \begin{vmatrix} \mathbf{i} & \mathbf{j} & \mathbf{k} \\ -1 & \cos\theta & \sin\theta \\ 0 & -r\sin\theta & r\cos\theta \end{vmatrix} \;=\; r\mathbf{i}+r\cos\theta\mathbf{j}+r\sin\theta\mathbf{k}.$$

$$\iint_S (\boldsymbol{\nabla}\times\boldsymbol{f})\cdot\boldsymbol{N}\,\mathrm{d}S \;=\; \iint_D (\boldsymbol{\nabla}\times\boldsymbol{f})\cdot\left(\frac{\partial\boldsymbol{x}}{\partial r}\times\frac{\partial\boldsymbol{x}}{\partial\theta}\right)\mathrm{d}r\,\mathrm{d}\theta \quad\text{(see Page 268)}$$

$$= \int_0^\pi\int_0^1 \left(1-r-2, 1-r-r\cos\theta, r^2\cos^2\theta\right)\cdot\left(r, r\cos\theta, r\sin\theta\right)\mathrm{d}r\,\mathrm{d}\theta$$

$$= \int_0^\pi\int_0^1 \left(-r(1+r)+r\cos\theta(1-r-r\cos\theta)+r^3\cos^2\theta\sin\theta\right)\mathrm{d}r\,\mathrm{d}\theta$$

$$= \int_0^\pi \left(-\frac{5}{6}+\frac{1}{6}\cos\theta-\frac{1}{3}\cos^2\theta+\frac{1}{4}\cos^2\theta\sin\theta\right)\mathrm{d}\theta.$$

So,

$$\iint_S (\boldsymbol{\nabla}\times\boldsymbol{f})\cdot\boldsymbol{N}\,\mathrm{d}S \;=\; -\frac{5\pi}{6}+0-\frac{1}{3}\cdot\frac{1}{2}\cdot\pi+\frac{1}{4}\cdot\frac{1}{3}\cdot2 \;=\; \frac{1}{6}-\pi,$$

and the theorem is verified.

The reader should redo the surface integral calculation using the piecewise planar combination alternative.

Be mindful of choosing the correct surface orientation.

■

### What is good about Stokes's theorem:

Assuming that the conditions of the theorem are satisfied, if we were asked to evaluate the work $\oint_\Gamma \boldsymbol{f} \cdot d\boldsymbol{r}$ around a closed contour, we could instead evaluate $\iint_S (\boldsymbol{\nabla} \times \boldsymbol{f}) \cdot d\boldsymbol{S}$ through any convenient surface that has $\Gamma$ as its boundary.

Alternatively, if we were asked to evaluate $\iint_S (\boldsymbol{\nabla} \times \boldsymbol{f}) \cdot d\boldsymbol{S}$ through a given $S$, we could instead evaluate the flux integral through any surface that has the same curve $\Gamma$ as its boundary.              — there is an infinity of choices
provided $\boldsymbol{f}$ is not singular

Or, we could evaluate $\oint_\Gamma \boldsymbol{f} \cdot d\boldsymbol{r}$, if $\boldsymbol{f}$ could be deduced from the given expression $\boldsymbol{\nabla} \times \boldsymbol{f}$.

In these cases we choose whichever integral is easiest to evaluate. (See Mastery Check 5.25.)

**What is NOT so good about Stokes's theorem:**

The conditions on the theorem are strict. As given, the theorem states that

$$\oint_\Gamma \boldsymbol{f} \cdot d\boldsymbol{r} = \iint_S (\boldsymbol{\nabla} \times \boldsymbol{f}) \cdot d\boldsymbol{S}$$

is true *only if*

(a) $\Gamma$ is a closed contour. Unfortunately, not every problem posed involves a closed curve. However, if the given curve $\Gamma$ is not closed, we make up a convenient closed contour by complementing $\Gamma$:

$$\Gamma \longrightarrow \Gamma_c = \Gamma \cup \Gamma_{\mathrm{xtra}}$$

closed        extra bit

We can then apply Stokes's theorem on $\Gamma_c$ since, by Corollary 5.2.1, it is reasonable to argue that

$$\int_\Gamma \boldsymbol{f} \cdot d\boldsymbol{r} = \oint_{\Gamma_c} \boldsymbol{f} \cdot d\boldsymbol{r} - \int_{\Gamma_{\mathrm{xtra}}} \boldsymbol{f} \cdot d\boldsymbol{r} = \iint_S (\boldsymbol{\nabla} \times \boldsymbol{f}) \cdot d\boldsymbol{S} - \int_{\Gamma_{\mathrm{xtra}}} \boldsymbol{f} \cdot d\boldsymbol{r}.$$

— by Stokes's theorem on $\Gamma_c$
at the middle step

(b) $f$ is non-singular on $S$ or $\Gamma$. If $f$ is singular somewhere, then it is harder to deal with, and success will depend on the problem.

On the other hand, if we are asked to evaluate $\oint_\Gamma f \cdot dr$, and $f$ is singular on the given $S$, we can introduce an extra closed contour $\Gamma_\epsilon$ of small size $\epsilon$ around the singular point on $S$ so as to exclude it: $\Gamma \longrightarrow \Gamma_c = \Gamma \cup \Gamma_\epsilon$. Introducing this new contour on $S$ defines a new surface $S_c$ bounded by composite boundary $\Gamma_c$. Hence, with the singular point now removed we can argue, again by appealing to Corollary 5.2.1, that

$$\oint_\Gamma f \cdot dr = \oint_{\Gamma_c} f \cdot dr - \oint_{\Gamma_\epsilon} f \cdot dr = \iint_{S_c} \nabla \times f \cdot N \, dS - \oint_{\Gamma_\epsilon} f \cdot dr.$$

— by Stokes's theorem on $\Gamma_c$
at the middle step

We then need to consider the contribution from the curve integral around $\Gamma_\epsilon$ in the limit $\epsilon \to 0$ to regain the original surface $S$.

✍  **Mastery Check 5.25:**

Determine $\oint_\Gamma f \cdot dr$, where

$f = (y, x, xy)$ and $\Gamma$ is the curve of intersection of the plane $x + y + z = 1$ and

planes $x = y = z = 0$ in Figure 5.56.

Orientation of $\Gamma$: counterclockwise seen from $(10, 10, 10)$.

✍

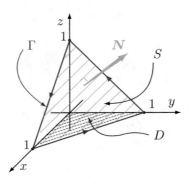

**Figure 5.56** The intersection of $x + y + z = 1$ and the coordinate planes.

# 5.G   Supplementary problems

### Section 5.A

1. Given a $C^2$ curve in $\mathbb{R}^3$ representing the trajectory of a particle and described by the parametrization

$$\boldsymbol{r}(t) = x(t)\mathbf{i} + y(t)\mathbf{j} + z(t)\mathbf{k}, \ 0 \le t \le T,$$

express the position, velocity and acceleration of the particle in terms of the local unit vectors $\boldsymbol{T}$, $\boldsymbol{N}$ and $\boldsymbol{B}$, the curvature $\kappa$ and torsion $\tau$ of the curve, and the speed $v$ of the particle. Give an interpretation of your result for the particle's acceleration.
   Conversely, express $\kappa$ and $\tau$ in terms of $\boldsymbol{v}$ and $\boldsymbol{a}$.

2. Find the curvature and torsion of the following smooth curves:

   (a) $\boldsymbol{r}(t) = (t^3, t^2, t)$, when $t = 1$,

   (b) $\boldsymbol{r}(t) = (\mathrm{e}^t, 3t, \mathrm{e}^{-2t})$, when $t = 1$,

   (c) $\boldsymbol{r}(t) = (t, t^2, \frac{2}{3}t^3)$, for general $t$.

3. Suppose a smooth curve is to be parameterized specifically in terms of arc length $s$ rather than $t$. From the equations derived earlier, express $\boldsymbol{T}$, $\boldsymbol{N}$ and $\boldsymbol{B}$ in terms of $s$. Consequently, derive formulae for $\boldsymbol{T}'(s)$, $\boldsymbol{N}'(s)$ and $\boldsymbol{B}'(s)$ corresponding to those for $\boldsymbol{T}'(t)$, $\boldsymbol{N}'(t)$ and $\boldsymbol{B}'(t)$ appearing in Definitions 5.4 and 5.5, and Mastery Check 5.4.
   The formulae for $\boldsymbol{T}'(s)$, $\boldsymbol{N}'(s)$ and $\boldsymbol{B}'(s)$ are called *Frenet-Serret* formulae and are fundamental to the differential geometry of space curves.

4. Consider the 3D circle of radius $a$ with centre at $\boldsymbol{r}_0$ described by

$$\mathbf{r}(\theta) = \mathbf{r}_0 + a(\sin\phi\cos\theta, \sin\phi\sin\theta, \cos\phi) \ \text{ with } \ \phi = f(\theta) \ \text{ and } \ \theta \in [0, 2\pi],$$

where $\theta$ and $\phi$ are spherical polar angle variables. Show that it has constant curvature, $\kappa(\theta) = 1/a$ and that the curve's torsion $\tau(\theta) = 0$.
   Hint: It may help to consider a few simple cases of circles.

   With this exercise we provide some detail to the claims made in Example 5.2.

5. The *catenary* curve in Figure 5.57(a) is the shape taken by a chain, of uniform density per unit length, which is allowed to hang under its

own weight.

Such a curve may be described by the vector function

$$\boldsymbol{r}(t) = at\boldsymbol{i} + a\cosh t\,\boldsymbol{j}, \quad -1 \le t \le 1,$$

where $a > 0$ is a constant.

Calculate the length of this curve.

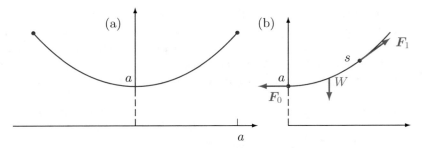

**Figure 5.57** (a) The curve $\boldsymbol{r}(t)\boldsymbol{i} + a\cosh t\boldsymbol{j}$ in Problem 4;
(b) The catenary curve in Problem 5.

6. Suppose a piece of chain is allowed to hang under its own weight,
   anchored at $(0, a)$, $a > 0$, by a force acting horizontally in the negative
   direction, and at some point $(x, y)$, $x > 0$, $y > a$, by a force tangential
   to the curve adopted by the chain.

   Let the piece of chain have length $s$ and let the curve be defined as
   $\boldsymbol{r}(t) = x(t)\,\boldsymbol{i} + y(t)\,\boldsymbol{j}$, $t \ge 0$. Assume the density of the chain is $\rho$ per
   unit length, and that it is at rest while subject to tangential tension
   forces at each end.

   These two tension forces $F_0$ and $F_1$ are shown in Figure 5.57(b).

   Assume (from elementary mechanics) that the horizontal components
   of these forces are equal and opposite, while the sum of the vertical
   components is equal to the weight $W = -\rho gs\boldsymbol{j}$ of that portion of the
   chain (acting through its centre of gravity).

   Show that if the forces are in equilibrium then the curve adopted by
   the chain is the catenary curve as defined in the last problem, with a
   suitable choice of constants. Use the following steps:

   (i) Show that $\dfrac{dy}{dx} = \dfrac{s}{\lambda}$ for some constant $\lambda$.

(ii) From $\dfrac{ds}{dt} = |\boldsymbol{r}'(t)|$ derive the result $\dfrac{ds}{dx} = \dfrac{\sqrt{\lambda^2 + s^2}}{\lambda}$, and a similar result for $\dfrac{ds}{dy}$.

(iii) Determine $x$ and $y$ in terms of $s$.

(iv) Show that these equations are solved by setting $x = \lambda t$, $y = \lambda \cosh t$. That is, the curve is as described with the appropriate choice of constants.

## Section 5.B

7. For smooth scalar functions $\psi$ and $\phi$, verify that
$$\boldsymbol{\nabla} \times (\phi \boldsymbol{\nabla} \psi) = \boldsymbol{\nabla}\phi \times \boldsymbol{\nabla}\psi.$$

8. For the $C^1$ fields $\boldsymbol{F}$ and $\boldsymbol{G}$, verify that
$$\boldsymbol{\nabla}(\boldsymbol{F} \cdot \boldsymbol{G}) = (\boldsymbol{F} \cdot \boldsymbol{\nabla})\boldsymbol{G} + (\boldsymbol{G} \cdot \boldsymbol{\nabla})\boldsymbol{F} + \boldsymbol{F} \times (\boldsymbol{\nabla} \times \boldsymbol{G}) + \boldsymbol{G} \times (\boldsymbol{\nabla} \times \boldsymbol{F}).$$

9. For the $C^1$ field $\boldsymbol{F}$, verify that
$$\boldsymbol{\nabla} \times (\boldsymbol{F} \times \boldsymbol{r}) = \boldsymbol{F} - (\boldsymbol{\nabla} \cdot \boldsymbol{F})\boldsymbol{r} + \boldsymbol{\nabla}(\boldsymbol{F} \cdot \boldsymbol{r}) - \boldsymbol{r} \times (\boldsymbol{\nabla} \times \boldsymbol{F}).$$

10. For the $C^2$ field $\boldsymbol{F}$, verify that
$$\boldsymbol{\nabla} \times (\boldsymbol{\nabla} \times \boldsymbol{F}) = \boldsymbol{\nabla}(\boldsymbol{\nabla} \cdot \boldsymbol{F}) - \boldsymbol{\nabla}^2 \boldsymbol{F}.$$

11. Show that in 3D a field proportional to $\dfrac{\boldsymbol{r}}{|\boldsymbol{r}|^3}$ is conservative. Show also that this field is solenoidal.

12. Show that the 2D electrostatic field $\boldsymbol{E} = \dfrac{\rho}{2\pi\epsilon_0 |\boldsymbol{r}|^2}\,\boldsymbol{r}$ (a field due to a uniformly charged wire of infinite length) is conservative.

13. Derive and solve the equations for the field lines corresponding to the 3D vector field $\dfrac{\boldsymbol{r}}{|\boldsymbol{r}|^3}$, $\boldsymbol{r} \neq \boldsymbol{0}$. Hence, show that these correspond to radial lines emanating from the origin.

14. Suppose the $C^1$ vector field $\boldsymbol{F}$ satisfies $\boldsymbol{\nabla} \cdot \boldsymbol{F} = 0$ in a domain $D \subset \mathbb{R}^3$. If
$$\boldsymbol{G}(x, y, z) = \int_0^1 t\boldsymbol{F}\left(\boldsymbol{r}(t)\right) \times \frac{d\boldsymbol{r}}{dt}\,dt$$

with $r(t) = (xt, yt, zt)$ for $t \in [0, 1]$, show that $\nabla \times G = F$. This shows that $G$ is a vector potential for the solenoidal field $F$ for points $(x, y, z) \in D$.

**Section 5.C**

15. Evaluate
$$\int_{(0,2)}^{(4,0)} (x^2 + y^2) \, dx$$
    along the path $y = \sqrt{4 - x}$.

16. Evaluate
$$\int_{(0,2)}^{(4,0)} (x^2 + y^2) \, dr$$
    along the path $y = \sqrt{4 - x}$.

17. Evaluate
$$\int_{(-1,0)}^{(1,0)} y(1 + x) \, dy$$
    a) along the $x$-axis; b) along the parabola $y = 1 - x^2$.

18. Along what curve of the family $y = kx(1 - x)$ does the integral
$$\int_{(0,0)}^{(1,0)} y(x - y) \, dx$$
    have the largest value?

19. Suppose a particle experiences a force directed to the origin with magnitude inversely proportional to its distance from the origin. Evaluate the work done by the force if the particle traverses the helical path $r(t) = (a \cos t, a \sin t, bt)$ for $t \in [0, 2\pi]$.

20. Compute $\int_{\Gamma} A \cdot dr$, where $A = -y\mathbf{i} + 2x\mathbf{j} + x\mathbf{k}$, and $\Gamma$ is a circle in the plane $x = y$ with centre $(1, 1, 0)$ and radius 1.

    Orientation: $\Gamma$ is traversed counterclockwise as seen from $(1, 0, 0)$, as shown in Figure 5.58.

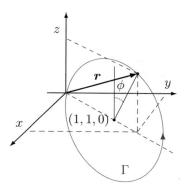

**Figure 5.58** A circle intersecting the $xy$-plane.

21. Compute $\int_\Gamma \boldsymbol{A} \cdot \mathrm{d}\boldsymbol{r}$, where

$$\boldsymbol{A} = x^2\mathbf{i} + (x^2 + y^2)\mathbf{j} + (x^2 + y^2 + z^2)\mathbf{k},$$

and $\Gamma$ is the boundary of $\{(x, y, z) : x + y + z = 1, \ x \geq 0, \ y \geq 0, \ z \geq 0\}$.

Orientation: $\Gamma$ is traversed counterclockwise as seen from the point $(10, 10, 10)$, as shown in Figure 5.56.

**Section** 5.D

22. Evaluate $\iint_S (2x + y - 3z^2)\,\mathrm{d}S$, where $S$ is defined by
$$\boldsymbol{r}(u, v) = u\,\mathbf{i} + v\,\mathbf{j} + u\,\mathbf{k}, \ 0 \leq u, v \leq 1.$$

23. Evaluate $\iint_S (6x + y - x^2)\,\mathrm{d}S$, where $S$ is defined by
$$\boldsymbol{r}(u, v) = u\,\mathbf{i} + u^2\,\mathbf{j} + v\,\mathbf{k}, \ 0 \leq u, v \leq 1.$$

24. Let $S$ be the ellipsoid

$$\frac{x^2}{a^2} + \frac{x^2}{b^2} + \frac{x^2}{c^2} = 1,$$

and $p(x, y, z)$ be the length of the perpendicular from the plane, that is tangent to $S$ at $(x, y, z) \in S$, to the origin. Show that the surface integral

$$\iint_S \frac{\mathrm{d}S}{p} = \frac{4}{3}\pi abc \left(\frac{1}{a^2} + \frac{1}{b^2} + \frac{1}{c^2}\right).$$

**Section** 5.E

25. Use Gauss's theorem to determine the flux of the vector field
$f = (x^2 + y^2, y^2 + z^2, z^2 + x^2)$ through the surface of the cone
$S = \{(x, y, z) : x^2 + y^2 - (z - 2)^2 = 0,\ 0 \le z \le 2,\ N \cdot e_3 > 0\}$ (Figure
5.59).

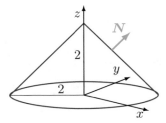

**Figure 5.59** The cone $x^2 + y^2 = (z - 2)^2$.

26. Suppose the region $D \subset \mathbb{R}^3$ is bounded by a closed surface $S$. Using
Gauss's theorem as a basis, prove the following variants of the theorem:

(a)
$$\iiint_D (\nabla \times F)\, \mathrm{d}V = \oiint_S (N \times F)\, \mathrm{d}S$$

(b)
$$\iiint_D (\nabla \phi)\, \mathrm{d}V = \oiint_S (\phi N)\, \mathrm{d}S$$

27. Let $A$ be the area of a region $D$ of the surface of a sphere of radius $R$
centred at the origin. Let $V$ be the volume of the solid cone comprising
all rays from the origin to points on $D$. Show that
$$V = \frac{1}{3} AR.$$

28. Show that the electric intensity due to a uniformly charged sphere at
points outside the sphere is the same as if the charge were concentrated
at the centre, while at points inside the sphere it is proportional to the
distance from the centre.

**Section** 5.F

29. Verify Green's theorem for the integral

$$\oint_\Gamma (x^2 + y)\,\mathrm{d}x - xy^2\,\mathrm{d}y$$

where $\Gamma$ is the boundary of the unit square with vertices (in order) $(0,0)$, $(1,0)$, $(1,1)$, $(0,1)$.

30. Verify Green's theorem for the integral

$$\oint_\Gamma (x - y)\,\mathrm{d}x + (x + y)\,\mathrm{d}y$$

where $\Gamma$, shown in Figure 5.60, is the boundary of the area in the first quadrant between the curves $y = x^2$ and $y^2 = x$ taken anticlockwise.

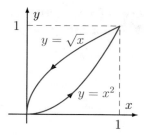

**Figure 5.60**  The closed contour $\Gamma$.

31. Verify Green's theorem for the integral

$$\oint_\Gamma (x - 2y)\,\mathrm{d}x + x\,\mathrm{d}y$$

where $\Gamma$ is the boundary of the unit circle $x^2 + y^2 = 1$ taken anticlockwise.

32. Use Green's theorem to evaluate

$$\oint_\Gamma (2xy + y^2)\,\mathrm{d}x + (x^2 + xy)x\,\mathrm{d}y$$

where $\Gamma$ is the boundary taken anticlockwise of the region cutoff from the first quadrant by the curve $y^2 = 1 - x^3$.

33. The intersection of the ellipsoid $x^2/2 + y^2 + (z - 1)^2/4 = 1$ and the plane $z + 2x = 2$ is a closed curve. Let $\Gamma$ be that part of the curve

lying above the $xy$-plane directed from, and including, $(1, -\frac{1}{2}, 0)$ to, and including, $(1, \frac{1}{2}, 0)$. Evaluate $\oint_\Gamma \boldsymbol{F} \cdot d\boldsymbol{r}$ if

$$\boldsymbol{F}(x, y, z) = \left(z^2 + \frac{2}{3}xy^3\right)\mathbf{i} + \left(x^2y^2\right)\mathbf{j} + \left(\frac{1}{4}z^2y^4 + \frac{1}{2}y^2\right)\mathbf{k}.$$

34. Let $\Gamma$ be the curve of intersection of the surfaces
$$S_1 = \{(x, y, z) : 4x^2 + 4y^2 + z^2 = 40\},$$
$$S_2 = \{(x, y, z) : x^2 + y^2 - z^2 = 0, z > 0\}.$$

Evaluate $\oint_\Gamma \boldsymbol{F} \cdot d\boldsymbol{r}$ if

$$\boldsymbol{F}(x, y, z) = \left(\frac{y}{x^2 + y^2} - z\right)\mathbf{i} + \left(x - \frac{x}{x^2 + y^2}\right)\mathbf{j} + \left(y + x^2\right)\mathbf{k}.$$

35. Suppose $\Gamma$ is the curve of intersection of surfaces $x^2 + y^2 = x$ and $1 - x^2 - y^2 = z$, while vector field $\boldsymbol{f} = (y, 1, x)$. Evaluate $\oint_\Gamma \boldsymbol{f} \cdot d\boldsymbol{r}$.

# Glossary of symbols

| | |
|---|---|
| $a, b, c, a_1, ..., b_1, ...$ | Scalar constants (usually). |
| $x, y, z, u, v, w, s, t$ | Variables, assumed continuous. |
| $\boldsymbol{u}, \boldsymbol{v}, \underline{u}, \underline{v}, \vec{u}, \vec{v}$ | Vector variables or vector constants. |
| $f, g, h, F, G, H$ | Functions (usually). |
| $F^{(k)}(x)$ | The $k^{\text{th}}$-order derivative of $F(x)$. |
| $m, n, \ell$ | Discrete integer variables. |
| $\mathbb{R}$ | Real one-dimensional space; the real line. |
| $\mathbb{R}^n,\ n = 2, 3, \ldots$ | Real $n$-dimensional space. $\mathbb{R}^3$ is the 3-dimensional space we inhabit. |
| $V, S$ | A volume region, and a surface embedded, in $\mathbb{R}^n$, $n \geq 3$. |
| $D_f;\ D,\ R$ | Domain of a function $f$; a general region and a rectangular region, respectively, (of integration) in $\mathbb{R}^n$. |
| $M$ | A subset of $\mathbb{R}^n$; $n$ is usually specified in context. |
| $M^c, \partial M$ | The complement, and the boundary, of a set $M$. |
| $\overline{M}$ | The closure of a set $M$: $\overline{M} = M \cup \partial M$. |
| $\boldsymbol{N}, \boldsymbol{n}$ | Unit normal vector and non-unit normal vector to a surface $S \subset \mathbb{R}^3$. |
| $|x|, |\boldsymbol{u}|$ | Absolute value of scalar $x$, and the magnitude of vector $\boldsymbol{u}$. |
| $|A|$ | Determinant of square matrix $A$. |

© Springer Nature Switzerland AG 2020
S. J. Miklavcic, An Illustrative Guide to Multivariable and Vector Calculus,
https://doi.org/10.1007/978-3-030-33459-8

| | |
|---|---|
| $\mathbf{i}, \mathbf{j}, \mathbf{k}$ | Unit vectors in the $x$-, $y$-, & $z$-directions, respectively. |
| $\mathbf{e}_1, \mathbf{e}_2, \mathbf{e}_3$ | Another way of writing $\mathbf{i}$, $\mathbf{j}$, $\mathbf{k}$, respectively. |
| $\theta, \phi$ | Angles (usually). The symbol $\phi$ is also used to denote a scalar potential field in $\mathbb{R}^n$ $n \geq 2$. |
| $\alpha, \beta, \gamma$ | Components of a fixed vector (usually). Angles (sometimes). |
| $\equiv, \exists, \forall$ | "equivalent to", "there exists", and "for all". |
| $\in$ | "belongs to" or "is a member of". |
| $\subset$ | "is a subset of" or "is contained within". |
| $\cap$ | $A \cap B$ is the intersection of two sets $A$, $B$. |
| $\cup$ | $A \cup B$ is the union of two sets $A$, $B$. |
| $(\cdot)$ | An undefined, undeclared, or generic, argument of a function. |
| s.t. | "such that". |
| : | (Inside a set-builder like $\{x : x < 0\}$) "such that". |
| w.r.t. | "with respect to". |
| $\dfrac{\mathrm{d}f}{\mathrm{d}x} \equiv f'(x)$ | Notation for the total derivative of a function $f$ of $x$. |
| $(f \circ g)(x)$ | Composite function; equivalent to $f\big(g(x)\big)$. |
| $\dfrac{\partial f}{\partial x} \equiv f_x$ | The partial derivative w.r.t. $x$ of a function $f$ of two or more variables. |
| $C^n(\mathbb{R}^m)$ | The set of continuous functions defined on the space $\mathbb{R}^m$ ($m \geq 1$) having continuous derivatives of order up to & including $n$. |
| $\nabla$ | The vector differential operator, "del". |
| $\nabla f$ | The gradient of a scalar function $f$, "grad $f$". |
| $\nabla \cdot \boldsymbol{f}$ | The divergence of a vector function $\boldsymbol{f}$, "div $\boldsymbol{f}$". |
| $\nabla \times \boldsymbol{f}$ | The rotation vector of a vector field $f$, "curl $\boldsymbol{f}$". |
| $J$ | The Jacobian determinant (usually). |
| $L$ | A linear operator, a level set, a line, or a length, depending on context. |

| | |
|---|---|
| $\Longrightarrow$, $\Rightarrow$ | Implication: "this implies" or "this results in". |
| $\Longleftrightarrow$ | "if and only if" or "equivalent to". |
| $\perp$ | "is orthogonal to", "is perpendicular to", or "is at right-angles to". |
| $\parallel$ | "is parallel to". |
| $\ll$ | "is much less than". |
| $\gtrless$ | "is greater than and also less than". |
| $\longmapsto$, $\mapsto$ | $f : x \longmapsto y$ (or $f : x \mapsto y$) "function $f$ maps point $x$ to point $y$" (point mapping). |
| $\longrightarrow$, $\rightarrow$ | Context dependent: $f : A \longrightarrow B$ (or $f : A \rightarrow B$) "function $f$ maps from set $A$ into set $B$" (set mapping); $x \longrightarrow 0$ (or $x \rightarrow 0$) "$x$ converges to $0$"; "tends to". |
| Field | A scalar or vector function on $\mathbb{R}^n$, $n \geq 2$. |
| Theorem | A proposition that can be proved to be true. |
| Corollary | A result that follows immediately from a theorem. |
| 1D, 2D, 3D | "one dimension" or "one-dimensional", etc. |
| ODE | "ordinary differential equation". |
| PDE | "partial differential equation". |
| b.c. | "boundary condition". |
| b.v.p. | "boundary-value problem". |

# Bibliography

Any serious scientist, engineer or mathematician should be in possession of a decent personal library of reference books. Likewise, a university or college library worthy of its name should have a ready and sufficient store of text books. Although hard-copy books are becoming unfashionable (perhaps, at least, to be replaced by electronic literature — "eBooks"), it is important for the student to follow up on the material presented here by reading some of the more specialist books. The list given below is far from exhaustive, but these books do cover the areas we have discussed as well as being some of my favourites.

## On multivariable and vector calculus:

1. Adams, R.A., Calculus, A.: Complete Course, 5th edn. Addison-Wesley, Boston (2003)
2. Apostol, T.M.: Mathematical Analysis: A Modern Approach to Advanced Calculus. Addison-Wesley, Boston (1957)
3. Courant, R., Hilbert, D.: Methods of Mathematical Physics, vol. 1 & 2. Wiley-Interscience, Hoboken (1962)
4. Hardy, G.H.: A Course of Pure Mathematics, 9th edn. Cambridge University Press, Cambridge (1949)
5. Kaplan, W.: Advanced Calculus, 5th edn. Addison-Wesley, Boston (2003)
6. Grossman, S.I.: Calculus, 3rd edn. Academic, New York (1984)
7. Spiegel, M.R.: Schaum's Outline of Advanced Calculus. McGraw-Hill, New York (1974)
8. Spiegel, M.R.: Schaum's Outline of Vector Analysis. McGraw-Hill, New York (1974)

© Springer Nature Switzerland AG 2020                                         305
S. J. Miklavcic, An Illustrative Guide to Multivariable and Vector Calculus,
https://doi.org/10.1007/978-3-030-33459-8

# On the approximation of functions:

9. Edwards, R.E.: Fourier Series: A Modern Introduction, vol. 1 & 2. Holt, Rinehart & Winston, New York (1967)
10. Rice, J.R.: The Approximation of Functions, vol. 1 & 2. Addison-Wesley, Boston (1964)


# On partial differential equations:

11. Epstein, B.: Partial Differential Equations: An Introduction. McGraw-Hill, New York (1962)
12. Greenspan, D.: Introduction to Partial Differential Equations. McGraw-Hill, New York (1961)
13. Jeffreys, H., Jeffreys, B.: Methods of Mathematical Physics, 3rd edn. Cambridge University Press, Cambridge (1966)
14. Kreyszig, E.: Advanced: Engineering Mathematics, 7th edn. Wiley, New York (1982)
15. Morse, P.M., Feschbach, H.: Methods of Theoretical Physics, vol. 1 & 2. McGraw-Hill, New York (1953)


# On linear algebra:

16. Lipschutz, S., Lipson, M.L.: Schaum's Outline of Linear Algebra, 5th edn. McGraw-Hill, New York (2013)


# A little history:

17. Bell, E.T.: Men of Mathematics, vol. 1 & 2. Pelican, Kingston (1965)
18. Cannell, D.M.: George Green: Miller and Mathematician 1793–1841. City of Nottingham Arts Department (1988)
19. Dunningham, G.: Waldo Carl Friedrich Gauss: Titan of Science. Hafner (1955)
20. Newman, J.R.: The World of Mathematics, vol. 1–4. George Allen & Unwin, Crows Nest (1961)


# A little grammar:

21. Fowler, H.W.: A Dictionary of Modern English Usage, 2nd edn (Revised by Sir Ernest Gowers). Oxford University Press, Oxford (1968)
22. Taggart, C., Wines, J.A.: My Grammar and I (or should that be 'me'?). Michael O'Marra Books Ltd (2011)

# Index

© Springer Nature Switzerland AG 2020
S. J. Miklavcic, An Illustrative Guide to Multivariable and Vector Calculus,
https://doi.org/10.1007/978-3-030-33459-8

Printed in the United States
By Bookmasters